AF790808

Nuclear Dynamics in Strong Laser Fields

Online at: https://doi.org/10.1088/978-0-7503-3199-9

Nuclear Dynamics in Strong Laser Fields

Şerban Mişicu
Department for Theoretical Physics, Horia Hulubei National Institute for R&D in Physics and Nuclear Engineering, Bucharest, Romania

IOP Publishing, Bristol, UK

© IOP Publishing Ltd 2023

All rights reserved. No part of this publication may be reproduced, stored in a retrieval system or transmitted in any form or by any means, electronic, mechanical, photocopying, recording or otherwise, without the prior permission of the publisher, or as expressly permitted by law or under terms agreed with the appropriate rights organization. Multiple copying is permitted in accordance with the terms of licences issued by the Copyright Licensing Agency, the Copyright Clearance Centre and other reproduction rights organizations.

Permission to make use of IOP Publishing content other than as set out above may be sought at permissions@ioppublishing.org.

Şerban Mişicu has asserted his right to be identified as the author of this work in accordance with sections 77 and 78 of the Copyright, Designs and Patents Act 1988.

ISBN 978-0-7503-3199-9 (ebook)
ISBN 978-0-7503-3197-5 (print)
ISBN 978-0-7503-3200-2 (myPrint)
ISBN 978-0-7503-3198-2 (mobi)

DOI 10.1088/978-0-7503-3199-9

Version: 20230801

IOP ebooks

British Library Cataloguing-in-Publication Data: A catalogue record for this book is available from the British Library.

Published by IOP Publishing, wholly owned by The Institute of Physics, London

IOP Publishing, No.2 The Distillery, Glassfields, Avon Street, Bristol, BS2 0GR, UK

US Office: IOP Publishing, Inc., 190 North Independence Mall West, Suite 601, Philadelphia, PA 19106, USA

Contents

Preface **vii**

Acknowledgement **ix**

Author biography **x**

List of symbols **xi**

1 Introduction **1-1**

1.1 Statement of the problem 1-1

1.2 A brief review of electromagnetic theory 1-2

1.3 Laser pulses 1-5

1.4 The motion of a charged particle in an electromagnetic field 1-6

1.4.1 The Lagrangian and Hamiltonian of a charged particle coupled to the electromagnetic field 1-6

1.5 Gauge transformations 1-9

1.5.1 The dipole approximation 1-10

1.5.2 The Göppert-Mayer (length) gauge 1-11

1.5.3 The velocity gauge 1-12

1.5.4 The Henneberger gauge 1-13

1.5.5 The Volkov wavefunction 1-15

1.6 The dinuclear system in a laser field 1-15

1.7 Characterization of laser–nucleus interaction in analogy to atom–nucleus interaction 1-18

1.8 Doppler amplification of the laser pulse characteristics 1-21

References 1-23

2 Theoretical treatment of classical and quantum systems in strong laser fields **2-1**

2.1 Oscillations of a classical plasma induced by a periodic electric field 2-1

2.2 Perturbative methods 2-4

2.2.1 Nonstationary perturbation theory 2-4

2.2.2 Perturbations periodic in time 2-6

2.2.3 Adiabatic perturbation 2-8

2.3 Non-perturbative methods 2-9

2.3.1 Exact treatment of the harmonic oscillator in a pulsed electromagnetic field 2-9

2.3.2 The quasi-energy method 2-12

2.3.3 Kramers–Henneberger method 2-15
2.4 High-intensity regimes 2-17
2.4.1 The strong-field approximation 2-17
References 2-24

3 Giant dipole modes induced by a high-frequency electric field 3-1

3.1 Neutron–proton mixtures in an external electromagnetic field 3-1
References 3-12

4 Laser-assisted nuclear radioactive processes 4-1

4.1 The semiclassical theory of quantum tunneling in a periodic field 4-1
4.2 The modification of nuclear particle decay rates in x-ray laser fields 4-4
4.2.1 Laser-assisted alpha-decay in a one-dimensional approach 4-4
4.2.2 Laser-assisted proton decay in a two-dimensional approach 4-14
4.2.3 Some cases of proton decay in spherical and deformed nuclei 4-21
References 4-28

5 Heavy-ion scattering in a strong laser field 5-1

5.1 Scattering in an arbitrarily polarized radiation field 5-2
5.2 Refractive scattering of loosely bound nuclei 5-7
5.3 Elastic scattering in a linearly polarized short laser pulse 5-13
References 5-16

Appendix A A-1

Preface

The interaction of strong laser pulses with atomic nuclei is a new and growing topic of fundamental research. While atomic processes are well suited to being studied using visible or near visible laser radiation, nuclear processes are accessible to light in the x- and gamma-ray domains. The efforts in this field have been stimulated by the worldwide advent of very powerful laser facilities. One of these, ELI-NP, located at Măgurele, in the south-eastern outskirts of Bucharest, Romania, is focusing on laser-based nuclear physics and will generate radiation and particle beams with higher energy and with brilliance suited to the study of fundamental processes at the interfaces of plasma, nuclear, and particle physics. In addition to a wide range of fundamental physics projects, a variety of applied research will also be carried out at ELI-NP. Among these, one should mention the development of techniques for remote characterization of nuclear materials or radioactive waste using nuclear resonance fluorescence. Over the last 20 years, a consistent number of articles discussing various problems of low-energy nuclear physics processes influenced by strong lasers from a theoretical perspective have appeared in high-ranking journals. The modification of α-decay rates under the action of a monochromatic, linearly polarized laser field and the possibility to accelerate the disposal of radioactive waste have attracted the attention of many authors. In addition to this theme I also mention the contributions dedicated to various laser-assisted nuclear processes such as the Stark effect, nuclear photoeffect, proton-decay, heavy-ion scattering, nuclear fusion, nuclear fission, stability of nuclear molecules, decay of nuclear isomeric states, and excitation of giant resonances.

These recent theoretical developments, as well the growing number of specialists from nuclear physics interested in studying the role played by ultra-intense laser pulses in initiating new processes pertaining to nuclear structure and nuclear reaction physics, presented a good motivation to write this book. One of the tasks of this endeavor was to present theoretical techniques, some of them borrowed from atomic physics, and applications related to important aspects of nuclear physics.

In chapter 1 the state-of-the-art of laser facilities is reviewed, followed by a brief review of the electromagnetic theory (the Maxwell equations, the motion of a charged particle in an electromagnetic field) and properties of a laser pulse. Some of the gauge transformations used to describe quantum states interacting with a radiation field are introduced and then discussed in the context of a dinuclear system in a laser field. Some of the most well-known parameters, originally employed to analyze the atomic effects induced by a laser field, are applied to characterize the nuclear effects.

The scope of chapter 2 is to present the theoretical methods used in the description of classical and quantum systems subjected to laser pulses that, although originally developed for classical plasma and atoms, can be applied also to atomic nuclei.

Chapter 3 describes the excitation of the giant dipole resonance in a classical proton–neutron fluid by a coherent pulse of high-energy photons from the γ-ray domain.

Chapter 4 is dedicated to the investigation of laser-assisted nuclear radioactive processes (α-, proton-decay) and the estimation of the corresponding decay rates using the non-perturbative approach of solving the time-dependent Schrödinger equation exactly.

Chapter 5 deals with the distortion of heavy-ion elastic scattering differential cross-sections in strong laser fields in the optical domain as well the enhancement of multi-photon effects in modulated laser pulses.

At the end of the book a mathematical appendix presents the classical motion of a charged particle in a high-frequency field and the main properties of the Bessel functions used in applications.

The book has emerged from the author's contributions to this new developing field during the last ten years and is intended for the benefit of researchers and graduate students interested in the exploration of atomic nuclei with strong laser pulses.

Acknowledgement

The author is grateful to Mrs Matei Consuela for reading the manuscript and formulating useful suggestions.

Author biography

Şerban Mişicu

Dr habil. Şerban Mişicu is a senior researcher in the Department of Theoretical Physics of the Horia Hulubei National Institute of Physics and Nuclear Engineering (Bucharest, Romania).

He received his PhD from the Institute of Atomic Physics (Bucharest-Magurele) in 1995 and the Habilitation from Johann Wolfgang Goethe University Frankfurt am Main.

He was an Alexander von Humboldt Fellow from 2000 to 2001 at the Institut für Theoretische Physik Frankfurt am Main, in 2005 at the Institut für Kernphysik Darmstadt and 2012 at Frankfurt Institute for Advanced Science. He received a Marie Curie Postdoctoral Fellowship from 2001 to 2003 at the Institut für Theoretische Physik Frankfurt am Main. He was also a Fulbright scholar from 2005 to 2006 at Argonne National Laboratory and received DAAD fellowships in 1999 and 2013

He is currently involved in theoretical research on the interaction of nuclei with strong laser pulses, nuclear collective motion, and heavy-ion reactions.

List of symbols

$\overrightarrow{\alpha}(t)$	quiver vector (m)
ε_0	electric permittivity in a vacuum ($C^2\ N^{-1}\ m^{-2}$)
Γ	spreading width of the giant resonance (s^{-1})
λ	wavelength (m)
μ_0	magnetic permittivity in a vacuum ($H\ m^{-1}$)
ω	circular frequency of an electromagnetic wave (s^{-1})
ω_p	plasma frequency (s^{-1})
$\mathbf{\Phi}(\mathbf{r}, t)$	scalar potential ($J\ C^{-1}$)
$\rho_{p(n)}$	proton(neutron) matter density ($kg\ m^{-3}$)
$\rho(\mathbf{r}, t)$	charge density ($C\ m^{-3}$)
$\chi(\mathbf{r}, t)$	gauge scalar function
A	nuclear mass
$\mathbf{A}(\mathbf{r},t)$	vector potential ($T\ m^{-1}$)
$\mathbf{B}(\mathbf{r},t)$	magnetic field (T)
$\mathbf{E}(\mathbf{r},t)$	electric field ($V\ m^{-1}$)
$\mathscr{E}_0$	strength of the electric field ($V\ m^{-1}$)
e	proton charge (C)
c	speed of light in a vacuum ($m\ s^{-1}$)
$\mathbf{d}$	dipole moment (C m)
I	intensity or time-average of the Poynting vector ($W\ cm^{-2}$)
$\mathbf{j}(\mathbf{r},t)$	current density ($C\ m^{-2}\ s^{-1}$)
m	mass of a particle (kg)
P_{tun}	tunneling probability
q	charge of a particle (C)
R_N	nuclear radius (m)
$\mathbf{S}(\mathbf{r}, t)$	radiated energy per unit time across a unit area (Poynting vector) ($J\ s^{-1}m^{-2}$)
$\mathbf{v}$	velocity of a particle ($m\ s^{-1}$)
u	energy density of the electromagnetic radiation ($J\ m^{-3}$)
U_p	ponderomotive energy (J)
Z	nuclear charge

Chapter 1

Introduction

1.1 Statement of the problem

In very recent times laser fields over a wide range of intensities and frequencies have been made available in the form of short pulses. Due to the emergence of chirped pulse amplification (CPA) techniques [1], along with the development of high-fluence laser materials, the generation of intensities in excess of 10^{18} W cm^{-2} has been achieved [2]. Gérard Mourou (France) and Donna Strickland (USA) were awarded the Nobel Prize for Physics in 2018 for contributing in a significant manner to the conception of this revolutionary technique [3].

Until very recently the available laser intensities and photon wavelengths were still insufficient for exploring strongly bound nuclear matter. However the construction of a new type of large-scale laser facility, the Extreme Light Infrastructure (ELI), developed near Bucharest in the last decade [4], will make it possible to overcome the distance on the energy scale between the atomic and the nuclear levels. This joint European project was designed to produce ultra-intense and ultra-short laser pulses of wavelength $\lambda = 800$ nm (using a Ti:sapphire laser) as well γ radiation by scattering of light photons on high energy electrons. The laser system is equipped with two 10 PW laser arms that are able to reach intensities of 10^{23} W cm^{-2} and electrical fields of up to 10^{15} V m^{-1}. Among other fields, studies of nuclear structure (e.g. exotic nuclei, photo-fission) and astrophysics are projected to be carried out at this outstanding facility.

Due to the low photon energies and insufficiently high power of laser pulses, excepting perhaps the elastic scattering of heavy ions, the main parts of the nuclear structure and reaction processes are only weakly disturbed; many low-energy nuclear excitations involve energies in the range 1 keV–10 MeV, whereas laser photon energies, such as those produced with the Ti:sapphire laser, barely exceed 1 eV. For this reason the soon-to-enter-into-operation ELI-NP facility intends to use ultra-intense laser fields to produce very brilliant, intense γ beams of up to 19 MeV,

doi:10.1088/978-0-7503-3199-9ch1 © IOP Publishing Ltd 2023

0.1% bandwidth, and $10^{13}\,\gamma\ \mathrm{s}^{-1}$ by Compton back-scattering, which will allow the investigation of nuclear structure and nuclear reactions [5, 6].

There are also endeavours to generate intense x-ray fields with the help of free-electron lasers. Such systems are able to produce photon energies above 10 keV (see [7, 8]) and one of these facilities, the Linac Coherent Light Source (LCLS) [9], is capable of generating laser light at sub-angstrom wavelengths by combining a short-period undulator with an 8 GeV electron beam, the maximum power barely exceeding 10 GW with a pulse duration of 10 fs. Note that the shortest wavelength attained, i.e. 0.634 Å, corresponds to a photon energy of 19.6 keV. Another x-ray free-electron laser (XFEL) facility is the SwissFEL at the Paul Scherrer Institute [10]. It provides very intense and tightly focused beams of x-rays with pulses as short as 10 fs (approximately the duration of ten molecular vibration periods) and wavelengths in the range of 1–7 Å by the ARAMIS hard x-ray beamlines, and 6.5–50 Å by the ATHOS soft x-ray beamlines. For the soft x-ray beamlines, a pulse energy in excess of 1 mJ, a pulse duration of 30 fs, and a maximum bandwidth of 0.25% was obtained. With each shot being focused into a $100 \times 100\ \mathrm{nm}^2$ spot, a peak power below 1 TW and an intensity up to $10^{22}\ \mathrm{W\ cm}^{-2}$ is expected [10]. It has been pointed out that intense laser pulses are expected to reach power densities of over $10^{20}\ \mathrm{W\ cm}^{-2}$ using 9.9. keV photons [11]. It would be therefore beneficial to have at hand for nuclear physics studies both the ultra-high intensities produced by ELI and the high photon energies of XFEL.

It has been highlighted by a significative number of authors that high-intensity lasers, ranging from the optical to the x-ray domain, are likely to influence various nuclear processes, e.g. nuclear transitions by electronic [12, 13] or muonic transitions [14], proton emission from nuclei via the nuclear photoeffect in an intense laser field accompanied by a γ-ray [15], modification of α [16–21] and proton decay rates [22, 23], proton inelastic scattering on $^{12}\mathrm{C}$ in monochromatic [24] and bichromatic fields [25], modification of heavy-ion elastic scattering differential cross-sections [26, 27], the deuteron–triton fusion probability enhancement [28–30] or the bimodal structure of the photoabsorption cross section [31].

In this chapter a brief review of the electromagnetic theory and a discussion of some of the most relevant characteristics of the laser pulses are expounded.

1.2 A brief review of electromagnetic theory

The *Maxwell equations* relate the electric $\mathbf{E}(\mathbf{r}, t)$ and magnetic $\mathbf{B}(\mathbf{r}, t)$ fields to the charge $\rho(\mathbf{r}, t)$ and current $\mathbf{j}(\mathbf{r}, t)$ densities

$$\nabla \cdot \mathbf{E}(\mathbf{r}, t) = \frac{1}{\varepsilon_0}\rho(\mathbf{r}, t) \tag{1.1}$$

$$\nabla \cdot \mathbf{B}(\mathbf{r}, t) = 0 \tag{1.2}$$

$$\nabla \times \mathbf{E}(\mathbf{r}, t) = -\frac{\partial}{\partial t}\mathbf{B}(\mathbf{r}, t), \tag{1.3}$$

$$\nabla \times \mathbf{B}(\mathbf{r}, t) = \mu_0 \mathbf{j}(\mathbf{r}, t) + \frac{1}{c^2}\frac{\partial}{\partial t}\mathbf{E}(\mathbf{r}, t). \tag{1.4}$$

Equations (1.2) and (1.3) lead to alternative forms of the fields **E** and **B**:

$$\mathbf{E}(\mathbf{r}, t) = -\frac{\partial}{\partial t}\mathbf{A}(\mathbf{r}, t) - \nabla\Phi(\mathbf{r}, t) \tag{1.5}$$

$$\mathbf{B}(\mathbf{r}, t) = \nabla \times \mathbf{A}(\mathbf{r}, t), \tag{1.6}$$

where **A** is the *vector potential* and Φ the *scalar potential*. These two equations show that **E** and **B** are invariant to the *gauge transformation* [33]

$$\Phi(\mathbf{r}, t) \longrightarrow \Phi(\mathbf{r}, t) - \frac{\partial}{\partial t}\chi(\mathbf{r}, t), \tag{1.7}$$

$$\mathbf{A}(\mathbf{r}, t) \longrightarrow \mathbf{A}(\mathbf{r}, t) + \nabla\chi(\mathbf{r}, t), \tag{1.8}$$

where $\chi(\mathbf{r}, t)$ is an arbitrary scalar function. The gauge transformation determines a certain ambiguity of the scalar and vector potentials, which has no direct physical meaning, since a given pair of fields (**E**, **B**) can be described by various potentials. This ambiguity can be partly removed by choosing a gauge condition for $\nabla \cdot \mathbf{A}$. The precise form of the gauge transformation is derived by first substituting (1.5) and (1.6) in the Maxwell equations (1.1) and (1.4),

$$\nabla^2\Phi + \frac{\partial}{\partial t}(\nabla \cdot \mathbf{A}(\mathbf{r}, t)) = -\frac{\rho}{\varepsilon_0}, \tag{1.9}$$

$$\nabla^2\mathbf{A} - \frac{1}{c^2}\frac{\partial^2\mathbf{A}}{\partial t^2} - \nabla\left(\nabla \cdot \mathbf{A}(\mathbf{r}, t) + \frac{1}{c^2}\frac{\partial}{\partial t}\Phi(\mathbf{r}, t)\right) = -\mu_0\mathbf{j}. \tag{1.10}$$

It is obvious that these two partial differential equations for the potentials can be decoupled by appealing to the *Lorentz condition*:

$$\nabla \cdot \mathbf{A}(\mathbf{r}, t) + \frac{1}{c^2}\frac{\partial}{\partial t}\Phi(\mathbf{r}, t) = 0. \tag{1.11}$$

Indeed, in this gauge the scalar and vector electromagnetic potentials fulfill the non-homogeneous wave equations:

$$\nabla^2\Phi - \frac{1}{c^2}\frac{\partial^2\Phi}{\partial t^2} = -\frac{\rho}{\varepsilon_0}, \tag{1.12}$$

$$\nabla^2\mathbf{A} - \frac{1}{c^2}\frac{\partial^2\mathbf{A}}{\partial t^2} = -\mu_0\mathbf{j}. \tag{1.13}$$

The above equations can be solved by means of the Fourier transform technique, as illustrated below for the scalar potential. The solution for the vector potential will follow in a completely analogous manner.

The Fourier integrals of the charge density and scalar potential are defined as

$$\{\rho, \Phi\}(\mathbf{r}, t) = \int \mathrm{d}^3\mathbf{k} \int \mathrm{d}\omega \ \mathrm{e}^{\mathrm{i}(\mathbf{k}\cdot\mathbf{r}-\omega t)}\{\tilde{\rho}, \tilde{\Phi}\}(\mathbf{k}, \omega), \tag{1.14}$$

where the Fourier amplitudes are obtained via the inverse transformations

$$\{\tilde{\rho}, \tilde{\Phi}\}(\mathbf{k}, \omega) = \frac{1}{(2\pi)^4} \int \mathrm{d}^3\mathbf{r}' \int \mathrm{d}t' \ \mathrm{e}^{-\mathrm{i}(\mathbf{k}\cdot\mathbf{r}'-\omega t')}\{\rho, \Phi\}(\mathbf{r}', t'). \tag{1.15}$$

Inserting these expressions in the wave equation for the scalar field (1.12), a relation between these Fourier components is obtained,

$$\left(\mathbf{k}^2 - \frac{\omega^2}{c^2}\right)\tilde{\Phi} = \frac{\tilde{\rho}}{\varepsilon_0}. \tag{1.16}$$

Back substitution into equation (1.14) provides the scalar potential

$$\Phi(\mathbf{r}, t) = \frac{1}{16\pi^4\varepsilon_0} \int \mathrm{d}^3\mathbf{k} \int \mathrm{d}\omega \ \int \mathrm{d}^3\mathbf{r}' \int \mathrm{d}t' \ \frac{\rho(\mathbf{r}', t')\mathrm{e}^{\mathrm{i}\mathbf{k}\cdot(\mathbf{r}-\mathbf{r}')+\mathrm{i}\omega(t-t')}}{k^2 - \omega^2/c^2}. \tag{1.17}$$

The decoupling of the set (1.9)–(1.9) can be also achieved via the *Coulomb*, *radiation* or *transverse gauge*

$$\nabla \cdot \mathbf{A} = 0. \tag{1.18}$$

In this gauge

$$\nabla^2\Phi = -\frac{\rho}{\varepsilon_0}, \tag{1.19}$$

and

$$\nabla^2\mathbf{A} - \frac{1}{c^2}\frac{\partial^2\mathbf{A}}{\partial t^2} = -\mu_0\mathbf{j} + \frac{1}{c^2}\nabla\frac{\partial\Phi}{\partial t}. \tag{1.20}$$

The solution of the Poisson equation is given by

$$\Phi(\mathbf{r}, t) = \frac{1}{4\pi\varepsilon_0} \int \mathrm{d}\mathbf{r}' \ \frac{\rho(\mathbf{r}', t)}{|\mathbf{r} - \mathbf{r}'|}. \tag{1.21}$$

As an electromagnetic wave, the laser pulse carries energy and momentum. This content can be described by considering the so-called *radiant energy per volume*, or energy density. According to the Poynting theorem the electromagnetic energy flowing through space is shared equally between the electric and the magnetic components of the wave, i.e.

$$u = \frac{1}{2}\varepsilon_0\mathbf{E}^2 + \frac{1}{2\mu_0}\mathbf{B}^2. \tag{1.22}$$

The transport of electromagnetic energy per unit time across a unit area (measured in SI units W cm^{-2}) is described in isotropic media by the *Poynting vector*[1]

$$\mathbf{S} = \frac{1}{\mu_0}\mathbf{E} \times \mathbf{B}, \tag{1.23}$$

which is a rapidly varying function in time at optical frequencies and naturally the instantaneous value is hard to measure directly. Another quantity, which provides the average energy per unit area per unit time, is the time-averaged value of the Poynting vector magnitude

$$\langle S \rangle_T = \frac{1}{T}\int_{t-T/2}^{t+T/2} |\mathbf{S}(t)|\mathrm{d}t, \tag{1.24}$$

and assuming that the electromagnetic wave is linearly polarized and travels in the direction of the vector $\mathbf{k}$,

$$\{\mathbf{E},\, \mathbf{B}\} = \{\mathbf{E}_0,\, \mathbf{B}_0\}\cos(\mathbf{k}\cdot\mathbf{r} - \omega t) \tag{1.25}$$

it can be rewritten as

$$\langle S \rangle_T = \frac{c^2\varepsilon_0}{2}|\mathbf{E}_0 \times \mathbf{B}_0|, \tag{1.26}$$

or

$$I \equiv \langle S \rangle_T = \frac{c\varepsilon_0}{2}\mathcal{E}_0^2. \tag{1.27}$$

1.3 Laser pulses

Since the wavelength λ of an incoming light field, belonging to an optical or x-ray laser, is much longer than the nuclear size, the field can approximately be considered to be independent of the spatial coordinates. In this limit (the dipole approximation) the interaction between a charged nuclear particle and the external electromagnetic field is dominated by its electric component while the magnetic component is neglected. Therefore, in the dipole approximation the electric field of a non-dispersive laser pulse can be expressed by a modulated linear-polarized and monochromatic plane wave (single-mode field),

$$E(t) = \mathcal{E}_0 F(t)\sin(\omega t + \delta). \tag{1.28}$$

Above, $\mathcal{E}_0$ is the electric field strength, δ is the carrier-envelope phase (CEP), which is of no concern for long pulses, and $F(t)$ is the pulse shape function (envelope),

[1] John Henry Poynting (1852–1914)—English physicist, member of the Royal Society of London, and Professor at the University of Birmingham. He made important contributions to the physics of electrical phenomena, energy transport, radiation theory, and pressure of light. He introduced in 1884 the concept of electromagnetic energy flux (the Poynting vector), in 1891 measured the Earth's density, and in 1893 determined the constant of gravity.

normalized to one at its maximum, such that $\mathscr{E}_0$ is the peak of the pulse. Choosing $\delta = \pi/2$, the maximum of the laser pulse envelope occurs at $t = 0$. The half maximum of the full width of the intensity profile $|\mathbf{E}(t)|^2$ is defined as the pulse duration τ_p. Since laser pulses are very long compared with a single cycle, $\omega\tau_p \gg 1$. In what follows various types of envelopes are considered, some of them being frequently encountered in the literature [32], such as the sine-squared profile

$$F(t) = \sin^2\left(\frac{\pi t}{\tau_p}\right)\theta(\tau_p - t), \tag{1.29}$$

the Gaussian profile

$$F(t) = \exp\left[-4\pi^2 t^2/\tau_p^2\right]\theta(\tau_p - t), \tag{1.30}$$

the rectangular (step-wise function) profile

$$F(t) = \theta(\tau_p - t), \tag{1.31}$$

and the sequences of rectangular pulses

$$F(t) = \sum_{p=0}^{N}(-)^p\theta(t - \tau_p). \tag{1.32}$$

For this last profile the laser pulse acts in the time intervals $[\tau_0 = 0, \tau_1]$, $[\tau_2, \tau_3]$, etc. When $\tau_0 \longrightarrow \tau_p \gg T = 2\pi/\omega$, the signal coincides with a continuous wave (CW). In the general case short pulses with a constant amplitude of length equal to an integer number of half-cycles and separated by intervals of similar length are considered. This means that at time τ_{2i} the field is again turned on and then at τ_{2i+1} it is again turned off. Consequently the duration of the $(i + 1)$th short pulse is

$$\tau_{i+1} - \tau_i = n_{i+1}\frac{T}{2}, \tag{1.33}$$

where n_{i+1} is the number of half-cycles. The rectangular envelope is useful in analytical calculations.

For the applications discussed in this book the pulse duration is taken as an integer multiple of the laser period, $\tau_p = NT$. Routinely the minimum N is about a few dozens of cycles and in applications discussed in later chapters it is instructive to consider also pulse lengths of a few cycles.

1.4 The motion of a charged particle in an electromagnetic field

1.4.1 The Lagrangian and Hamiltonian of a charged particle coupled to the electromagnetic field

The motion of a particle of mass m and charge q, moving with velocity $\mathbf{v}$ in the electromagnetic field, specified by the intensities $\mathbf{E}$ and $\mathbf{B}$, is governed by the Lorentz force [33]

$$\mathbf{F}_{\text{Lorentz}} = m\dot{\mathbf{v}} = q(\mathbf{E} + \mathbf{v} \times \mathbf{B}). \tag{1.34}$$

In order to write the above equations in Lagrangian form and therefore to determine the corresponding Lagrange function, we make use of the previously defined electromagnetic potentials (1.5) and (1.6), and the Newton equation determined by the Lorentz forces reads

$$\frac{\mathrm{d}}{\mathrm{d}t}(m\dot{\mathbf{v}}) = q\left(-\nabla\Phi - \frac{\partial}{\partial t}\mathbf{A} + \mathbf{v} \times (\nabla \times \mathbf{A})\right). \tag{1.35}$$

Using the identity for the double vector product in the last term[2], we arrive at

$$\frac{\mathrm{d}}{\mathrm{d}t}(m\dot{\mathbf{v}}) = q\left(-\nabla\Phi - \frac{\partial}{\partial t}\mathbf{A} - (\mathbf{v} \cdot \nabla)\mathbf{A} + \nabla(\mathbf{v} \cdot \mathbf{A})\right). \tag{1.36}$$

Expressing the material derivative of $\mathbf{A}$ in terms of its covariant derivative, i.e.

$$\frac{\mathrm{d}}{\mathrm{d}t}\mathbf{A}(\mathbf{r}, t) = (\mathbf{v} \cdot \nabla)\mathbf{A} + \frac{\partial}{\partial t}\mathbf{A}, \tag{1.37}$$

the equation of motion (1.35) assumes the form

$$\frac{\mathrm{d}}{\mathrm{d}t}((m\dot{\mathbf{v}}) + q\mathbf{A}) = \nabla(-q\Phi + q\mathbf{v} \cdot \mathbf{A}). \tag{1.38}$$

The Lagrange equation satisfied by $L(\mathbf{r}, \mathbf{v}, t)$, written on components, reads [34]

$$\frac{\mathrm{d}}{\mathrm{d}t}\frac{\partial L}{\partial v_i} = \frac{\partial L}{\partial x_i}. \tag{1.39}$$

Introducing the canonical momentum

$$p_i = \frac{\partial L}{\partial \dot{x}_i} = m\dot{x}_i + qA_i, \tag{1.40}$$

and by comparing these last three expressions we are lead to the identification

$$\frac{\partial L}{\partial x_i} = \frac{\partial}{\partial x_i}(-q\Phi + q\mathbf{v} \cdot \mathbf{A}), \tag{1.41}$$

$$\frac{\partial L}{\partial \dot{x}_i} = m\dot{x}_i + qA_i. \tag{1.42}$$

From (1.41) it follows that

$$L = -q\Phi + q\mathbf{v} \cdot \mathbf{A} + f(\dot{\mathbf{r}}). \tag{1.43}$$

[2] $\mathbf{v} \times (\nabla \times \mathbf{A}) = \nabla(\mathbf{v} \cdot \mathbf{A}) - (\mathbf{v} \cdot \nabla)\mathbf{A}$ since $\mathbf{v}$ is not an explicit function of space.

Differentiating with respect to $\dot{x}_i$ and comparing with (1.42)

$$\frac{\partial f}{\partial \dot{x}_i} = m\dot{x}_i, \tag{1.44}$$

which upon integration leads to

$$f = \frac{1}{2}m\mathbf{v}^2. \tag{1.45}$$

Thus, the Lagrange equation for the non-relativistic motion of a charged particle in a given electromagnetic field reads

$$L(\mathbf{r}, \mathbf{v}, t) = \frac{1}{2}m\mathbf{v}^2 + q(\mathbf{v} \cdot \mathbf{A}(\mathbf{r}, t) - \Phi(\mathbf{r}, t)). \tag{1.46}$$

According to the Lagrange formulation of analytic mechanics, equation (1.42) provides the *canonical impulse*, composed out of the *mechanical impulse* $m\mathbf{v}$ and the electromagnetic impulse $q\mathbf{A}$,

$$\mathbf{p} = m\mathbf{v} + q\mathbf{A}. \tag{1.47}$$

Whereas the Lagrange formulation of analytic mechanics relies on the variables $(x_i, \dot{x}_i)$, the Hamilton theory uses the generalized position coordinate x_i and canonical impulse. The relation between the Lagrange and Hamilton functions for a particle is

$$H(\mathbf{r}, \mathbf{p}, t) = \mathbf{p} \cdot \mathbf{r} - L(\mathbf{r}, \mathbf{v}, t). \tag{1.48}$$

Thus, the Hamiltonian of a spinless particle of mass m, momentum $\mathbf{p}$, and charge q moving in a central potential $V(\mathbf{r})$ and coupled to an external electromagnetic field described by the potentials $\mathbf{A}$ and Φ, reads

$$H(\mathbf{r}, t) = \frac{1}{2m}[\mathbf{p} - q\mathbf{A}(\mathbf{r}, t)]^2 + V(\mathbf{r}) + q\Phi(\mathbf{r}, t). \tag{1.49}$$

The Hamilton function can be reformulated by expanding the square in the rhs of the above equation and separating the free part of the Hamiltonian from the interacting part

$$H = H_0 + H_{\text{int}}, \tag{1.50}$$

where

$$H_0 = \frac{1}{2m}\mathbf{p}^2 + V(\mathbf{r}). \tag{1.51}$$

and

$$H_{\text{int}} = -\frac{q}{m}\mathbf{p} \cdot \mathbf{A}(\mathbf{r}, t) + \frac{q^2}{2m}\mathbf{A}^2(\mathbf{r}, t) + q\Phi(\mathbf{r}, t). \tag{1.52}$$

1.5 Gauge transformations

A gauge transformation is a unitary transformation on the states of a quantum system. In classical physics gauge invariance signifies that physically measurable quantities preserve their values upon a gauge transformation of the electromagnetic quantities. In a more quantitative formulation it means that if an arbitrary function depending only on time is added to the Hamiltonian of the system, then the equations of motion are unchanged. Conversely, in quantum mechanics it is now a well-established fact that various gauges can lead to different physical interpretations despite the fact that the electromagnetic fields are invariant [35]. Reiss enumerated a series of common misconceptions related to gauge transformations in quantum mechanics and pointed out first of all that the interaction Hamiltonians do not transform unitarily, a fact that has the annoying consequence that a transition matrix element is not gauge invariant. From an experimental point of view this difference from classical physics can be explained by the fact that in quantum mechanics measurements are not done directly but rather by analyzing the data in terms of laboratory states that are not gauge invariant.

The form of the Hamiltonian of a quantum charged particle interacting simultaneously with the radiation field and a central potential (1.49) was derived earlier. Let us now generalize the previous considerations for a single charged particle to a system of Z identical particles of charge q (such as electrons in an atom or protons in a nucleus) specified by the coordinates $\mathbf{r}_\alpha$ and conjugate impulses $\mathbf{p}_\alpha = m\dot{\mathbf{r}}_\alpha + q\mathbf{A}(\mathbf{r}_\alpha, t)$, where α is an integer running from 1 to Z. Thus,

$$H = \sum_\alpha \left\{ \frac{1}{2m_\alpha}[\mathbf{p}_\alpha - q\mathbf{A}(\mathbf{r}_\alpha, t)]^2 + V(\mathbf{r}_\alpha) + q\Phi(\mathbf{r}_\alpha, t) \right\}. \tag{1.53}$$

As shown below, the Hamilton operator is changed by means of unitary transformations (gauge transformation). To insure the invariance of the wave equation associated with the above Hamiltonian, i.e.

$$\mathrm{i}\hbar\frac{\partial\Psi}{\partial t} = H\Psi, \tag{1.54}$$

the electromagnetic potentials ($\mathbf{A}$, Φ) are substituted in H according to (1.7) and (1.8). Consequently the wavefunction Ψ' satisfying the Schrödinger equation for the gauge transformed Hamiltonian

$$\mathrm{i}\hbar\frac{\partial\Psi'}{\partial t} = H\Psi', \tag{1.55}$$

can be obtained from Ψ by means of the phase transformation

$$\Psi' = \hat{U}\Psi, \quad \hat{U} = \exp\left[\frac{\mathrm{i}}{\hbar}q\chi(\mathbf{r}_\alpha, t)\right], \tag{1.56}$$

The invariance can be easily checked by observing that transformation (1.56) leaves the generalized momentum operator $\mathbf{p}_\alpha - q\mathbf{A}(\mathbf{r}_\alpha, t)$ unchanged, i.e.

$$\mathbf{p}'_\alpha - q\mathbf{A}'(\mathbf{r}_\alpha, t) = \hat{U}[\mathbf{p}_\alpha - q\mathbf{A}_\alpha(\mathbf{r}_\alpha, t)]\hat{U}^{-1}. \tag{1.57}$$

All quantities that are not changed by this transformation are said to be *gauge invariant*. Notorious examples are given by the probability density,

$$\rho = |\Psi|^2, \tag{1.58}$$

and the charge current density,

$$\begin{aligned}\mathbf{j} &= \frac{i\hbar q}{2m}\left[\Psi\left(\nabla + \frac{iq}{\hbar}\mathbf{A}\right)\Psi^* - \Psi^*\left(\nabla - \frac{iq}{\hbar}\mathbf{A}\right)\Psi\right] \\ &= \frac{i\hbar q}{2m}(\Psi\nabla\Psi^* - \Psi^*\nabla\Psi) - \frac{q^2}{m}\mathbf{A}|\Psi|^2.\end{aligned} \tag{1.59}$$

On the other hand the Hamiltonian (1.49) changes to

$$H' = \hat{U}H\hat{U}^{-1} - q\sum_\alpha \frac{\partial}{\partial t}\chi(\mathbf{r}_\alpha, t), \tag{1.60}$$

and therefore is no longer representing the same physical quantity once transformed via $\hat{U}$ since the eigenstates of this operator are different in the two representations (original and primed representations).

1.5.1 The dipole approximation

If the system of charges has a spatial extension R, small compared to the radiation wavelength $\lambda = 2\pi c/\omega$ (long wavelength approximation), the space part of the phase of the propagating plane wave, $\mathbf{k}\cdot\mathbf{r}$, is unimportant compared to the time part ωt,

$$\omega t - \mathbf{k}\cdot\mathbf{r} \approx \omega t. \tag{1.61}$$

Thus, it can be safely concluded that the vector potential is spatially homogeneous,

$$\mathbf{A}(\mathbf{r}_\alpha, t) \approx \mathbf{A}(0, t), \tag{1.62}$$

or, in other words, the lowest order term is retained in the expansion around the origin $\sum_\alpha \mathbf{r}_\alpha/Z = 0$, which is precisely the dipole term (*dipole approximation*),

$$H = \sum_\alpha \left\{\frac{1}{2m_\alpha}[\mathbf{p}_\alpha - q\mathbf{A}(0, t)]^2 + V(\mathbf{r}_\alpha) + \mathbf{d}\cdot\nabla\Phi(0, t)\right\}, \tag{1.63}$$

where

$$\mathbf{d} = q\sum_\alpha \mathbf{r}_\alpha \tag{1.64}$$

is the electric dipole moment. Note that the monopole term can be simply removed by a choice of coordinates, and since the wave number $k = 2\pi/\lambda = \omega/c$, the condition $R \ll \lambda$ shows that the long wavelength approximation sets an upper limit on the frequency [36], i.e.

$$\omega \ll \frac{2\pi c}{R}. \tag{1.65}$$

In addition to this limitation, Reiss pointed out [37] that within the dipole approximation there is also a lower limit for the frequency appearing due to a second-order effect which consists in the back-and-forth oscillation of the variable electric field induced by the magnetic field component of the laser field. In order to insure the validity of the dipole approximation, the onset of magnetic field effects must be avoided. For that purpose the amplitude parallel to the propagation direction,

$$\beta_0 = \frac{U_\mathrm{p}}{4\hbar\omega\left(1 + \dfrac{U_\mathrm{p}}{mc^2}\right)} \approx \frac{U_\mathrm{p}}{4\hbar\omega}, \tag{1.66}$$

a quantity that represents a direct measure of the magnetic field effect, should be small compared to one. Since the ponderomotive energy for a particle of mass m and charge q is given by

$$U_\mathrm{p} = \frac{q^2}{2\varepsilon_0 cm}\frac{I}{\omega^2}, \tag{1.67}$$

the lower limitation in frequency is expressed as

$$\omega^3 \gg \frac{q^2}{8\varepsilon_0 m\hbar c} I. \tag{1.68}$$

1.5.2 The Göppert-Mayer (length) gauge

The length gauge[3] is based on the approximation that the electric component of the laser field varies so slowly in space that it can be considered constant over the spatial extent R of the physical system and therefore the electromagnetic field appearing in the Hamiltonian can be described by a scalar field.

For a charged particle, the unitary transformation (1.56) defined by the Ansatz

$$\chi(\mathbf{r}, t) = -\mathbf{r} \cdot \mathbf{A}(0, t), \tag{1.69}$$

corresponds, according to equations (1.7) and (1.8), to the new potentials

$$\Phi'(\mathbf{r}, t) = \Phi(\mathbf{r}, t) + \mathbf{r} \cdot \dot{\mathbf{A}}(0, t), \tag{1.70}$$

$$\mathbf{A}'(\mathbf{r}, t) = \mathbf{A}(\mathbf{r}, t) - \mathbf{A}(0, t). \tag{1.71}$$

[3] This gauge was introduced by the German physicist Maria Göppert-Mayer (1906–72) in 1931. She is mostly known for her 1949 fundamental work on the nuclear shell model for which she was awarded the Nobel Prize in Physics in 1963 and also for the textbook on *Statistical Mechanics* co-authored with her husband Joseph E Mayer.

Thus,

$$\nabla\Phi'(0, t) = \nabla\Phi(0, t) + \dot{\mathbf{A}}(0, t) = -\mathbf{E}(\mathbf{r}, t), \tag{1.72}$$

$$\mathbf{A}'(0, t) = 0, \tag{1.73}$$

where in the rhs of (1.72) the alternative Maxwell form of the electric field (1.5) was used. Upon substitution of the last two identities in equation (1.63), the Hamiltonian generalized to the many-body case, is brought to a form where the electric dipole interaction is explicit,

$$H = \sum_{\alpha}\left\{\frac{1}{2m}\mathbf{p}_{\alpha}^{2} + V(\mathbf{r}_{\alpha}) - \mathbf{d}\cdot\mathbf{E}(0, t)\right\}. \tag{1.74}$$

Note that the generalized impulse is transformed in the conjugate impulse

$$\mathbf{p}_{\alpha}^{'} = m\dot{\mathbf{r}}_{\alpha}, \tag{1.75}$$

which shows that the Göppert-Mayer transformation restores the relation between the conjugate momentum of a particle, associated to its coordinate in the Lagrangian picture, and the velocity, by operating a change in the former quantity. In summary the unitary Göppert-Mayer transformation is

$$\hat{U}_{\mathrm{L}} = \exp\left[-\frac{\mathrm{i}}{\hbar}\mathbf{d}\cdot\mathbf{A}(0, t)\right]. \tag{1.76}$$

Above, the label L, commonly used in the literature, indicates the *length gauge*.

1.5.3 The velocity gauge

Another simplification of (1.49) is achieved by removing the term $\sim\mathbf{A}^2$ via a unitary transformation which singles out a time-dependent phase factor from the wavefunction

$$\chi(\mathbf{r}, t) = \frac{q}{2m}\int^{t}\mathrm{d}t'\mathbf{A}^2(0, t'). \tag{1.77}$$

The transformed potentials in this gauge are

$$\Phi'(\mathbf{r}, t) = \Phi(\mathbf{r}, t) - \frac{q}{m}\mathbf{A}(0, t)\cdot\dot{\mathbf{A}}(0, t), \tag{1.78}$$

$$\mathbf{A}'(\mathbf{r}, t) = \mathbf{A}(\mathbf{r}, t). \tag{1.79}$$

Thus,

$$\nabla\Phi'(\mathbf{r}, t) = \nabla\Phi(\mathbf{r}, t), \tag{1.80}$$

$$\mathbf{A}'(0, t) = \mathbf{A}(0, t). \tag{1.81}$$

The transformed Hamiltonian in the many-body case reads

$$H = \sum_{\alpha}\left[\frac{\mathbf{p}_{\alpha}^{2}}{2m} + V(\mathbf{r}_{\alpha}) - \frac{q}{m}\mathbf{p}_{\alpha}\cdot\mathbf{A}(0, t)\right]. \tag{1.82}$$

Since the interaction Hamiltonian, $q\mathbf{p}_\alpha \cdot \mathbf{A}/m$, couples the relative velocity to the vector potential, this gauge is dubbed the *velocity gauge*.

Reiss argued that the gauge transformation that removes the term $\sim\mathbf{A}^2$ is nonphysical and serves only as a mathematical simplification [35]. Since in current laser experiments $\mathbf{A}^2$ can be very large even when the dipole approximation is applicable, it seems that the elimination of the corresponding term in the Hamiltonian is justified. However, if a relativistic treatment is carried out, $\mathbf{A}^2$ is a function of $\mathbf{r}$ and t and therefore cannot be removed by a simple transformation. Reiss also pointed out that when the dipole approximation is valid, there is a gauge equivalence between the length and velocity gauges, which, however, fails at both short wavelengths and long wavelengths.

1.5.4 The Henneberger gauge

The transformation corresponding to this gauge has the goal to remove both terms in the interacting part of the Hamiltonian. The Kramers–Henneberger transformation, put forward in [38] and inspired from an earlier work of Kramers[4] [39], is defined by the scalar function

$$\chi(\mathbf{r}, t) = \frac{q}{m}\int_{-\infty}^{t} \mathrm{d}t' \left\{\mathbf{p} - \frac{1}{2}q\mathbf{A}(0, t')\right\} \cdot \mathbf{A}(0, t'). \tag{1.83}$$

It corresponds, according to (1.7) and (1.8), to the new potentials

$$\Phi'(\mathbf{r}, t) = \Phi(\mathbf{r}, t) - \frac{q}{m}\left\{\mathbf{p}(t) - \frac{1}{2}q\mathbf{A}(0, t)\right\} \cdot \mathbf{A}(0, t), \tag{1.84}$$

$$\mathbf{A}'(\mathbf{r}, t) = \mathbf{A}(\mathbf{r}, t). \tag{1.85}$$

Further

$$\nabla\Phi'(0, t) = 0, \tag{1.86}$$

$$\mathbf{A}'(\mathbf{r}, t) = \mathbf{A}(\mathbf{r}, t). \tag{1.87}$$

Thus,

$$\hat{U}_{\mathrm{K-H}} = \exp\left[\frac{\mathrm{i}}{\hbar}\frac{q^2}{2m}\int_{-\infty}^{t} \mathrm{d}t' H_{\mathrm{int}}(t')\right]. \tag{1.88}$$

The generalized momentum is unchanged by the K–H transformation,

[4] Hendrik Antonie Kramers (1894–1952) was a Dutch physicist, holder of the chair of theoretical physics at the University of Leiden. He was the first collaborator of Niels Bohr in the formulation and elaboration of the first modern theory of the atom. Many formulas and theories carry his name, e.g. the Kramers opacity formula, the Kramers–Kronig dispersion formula of classical electrodynamics, the Kramers dispersion theory, the WKB (Wentzel–Kramers–Brillouin) approximation method in quantum mechanics, the Kramers relation for chemical relations, etc. Kramers was the first to introduce the thermodynamic limit and derived, independently of Dirac, the relativistic wave equation for the electron and, independently of Debye, predicted the Compton effect.

$$\hat{U}_{\mathrm{K-H}}^{-1}\mathbf{p}'\hat{U}_{\mathrm{K-H}} = \mathbf{p}. \tag{1.89}$$

Instead, the arguments of the central potential are shifted by a time-dependent vector,

$$\hat{U}_{\mathrm{K-H}}^{-1}V(\mathbf{r})\hat{U}_{\mathrm{K-H}} = V(\mathbf{r} + \vec{\alpha}). \tag{1.90}$$

This new quantity was introduced by Henneberger [38] , using the short-hand notation,

$$\vec{\alpha}(t) = -\frac{q}{m}\int_{-\infty}^{t} \mathrm{d}t'\mathbf{A}(t'), \tag{1.91}$$

and corresponds to the displacement of the charged particle from its oscillation center in the electric field $\mathbf{E}(t)$, as can be inferred easily from the fact that it satisfies the classical equation of motion

$$\ddot{\vec{\alpha}} = -\frac{q}{m}\dot{\mathbf{A}}(t) = \frac{q}{m}\mathbf{E}(t). \tag{1.92}$$

In atomic physics $\vec{\alpha}$ describes a trajectory called *quiver* motion. It is related to the Hertz potential [39]

$$\vec{\alpha}(t) = \frac{q}{m}\mathbf{Z}(0,\, t'). \tag{1.93}$$

For a linear-polarized monochromatic plane wave (1.28) of amplitude,

$$\mathbf{A}_0 = \hat{\varepsilon}\frac{\mathscr{E}_0}{\omega}, \tag{1.94}$$

where $\hat{\varepsilon}$ is the polarization of the electromagnetic wave, the electric field, and the corresponding quiver vector in the dipole approximation and for $\delta = 0$ are written in the following form:

$$\mathbf{E}(t) = -\frac{\partial \mathbf{A}}{\partial t} = \hat{\varepsilon}\mathscr{E}_0 \sin \omega t, \tag{1.95}$$

$$\vec{\alpha}(t) = \hat{\varepsilon}\alpha_0 \sin \omega t, \tag{1.96}$$

where

$$\alpha_0 = -\frac{q\mathscr{E}_0}{m\omega^2}. \tag{1.97}$$

Thus, the many-body Hamiltonian in the oscillating Kramers–Henneberger frame is represented in the form

$$H = \sum_{\alpha}\left[\frac{\mathbf{p}_{\alpha}^2}{2m} + V(\mathbf{r} + \vec{\alpha}(t))\right]. \tag{1.98}$$

1.5.5 The Volkov wavefunction

The quantum dynamics of a charged particle, interacting solely with an electromagnetic wave, can be found by solving the Schrödinger equation (1.54) in the absence of the central field V,

$$i\hbar\frac{\partial}{\partial t}\Psi(\mathbf{r}, t) = \frac{1}{2m}[\mathbf{p} - q\mathbf{A}(\mathbf{r}, t)]^2\Psi(\mathbf{r}, t). \tag{1.99}$$

In the K–H representation (1.88) the wavefunction transforms to

$$\Psi_{\mathrm{K-H}} = \hat{U}_{\mathrm{K-H}}\Psi, \tag{1.100}$$

and the equation of motion (1.99) in the accelerated frame,

$$i\hbar\frac{\partial}{\partial t}\Psi_{\mathrm{K-H}} = \frac{1}{2m}\mathbf{p}^2\Psi_{\mathrm{K-H}}, \tag{1.101}$$

admits the solution

$$\Psi_{\mathrm{K-H}}(\mathbf{r}, t) = A\mathrm{e}^{\frac{\mathrm{i}}{\hbar}\left(\mathbf{p}\cdot\mathbf{r}-\frac{p^2}{2m}t\right)}. \tag{1.102}$$

Transforming back to the original representation,

$$\Psi(\mathbf{r}, t) = A\mathrm{e}^{\frac{\mathrm{i}}{\hbar}\left(\mathbf{p}\cdot\mathbf{r}-\frac{p^2}{2m}t-\vec{a}(t)\cdot\mathbf{p}-\frac{q^2}{2m}\int_{-\infty}^{t}\mathrm{d}t'\,\mathbf{A}^2(t')\right)}. \tag{1.103}$$

This is nothing other than the non-relativistic limit of the Volkov (Wolkow) wavefunction [40]

$$\Psi_V(\mathbf{r}, t) = A\mathrm{e}^{\frac{\mathrm{i}}{\hbar}\left(\mathbf{p}\cdot\mathbf{r}-\frac{1}{2m}\int_{-\infty}^{t}\mathrm{d}t'\,(\mathbf{p}-q\mathbf{A}(t'))^2\right)}. \tag{1.104}$$

1.6 The dinuclear system in a laser field

In what follows we set up the equations describing the quantum evolution of two charged nuclei, interacting via a central potential perturbed by an external electromagnetic field. The initial value problem for the Schrödinger equation associated to this system reads

$$i\hbar\frac{\partial\psi(\mathbf{r}_1, \mathbf{r}_2, t)}{\partial t} = H(t)\psi(\mathbf{r}_1, \mathbf{r}_2, t). \tag{1.105}$$

Consider as an example a dinuclear system consisting of a cluster (α) and a daughter (d) nucleus with masses $m_1 = m_\alpha$ and $m_2 = M_d = M - m_1$, charges $Z_1 = 2$, $Z_2 = Z - 2$, with positions in the laboratory frame specified by $\mathbf{r}_1$ and $\mathbf{r}_2$ and momenta $\mathbf{p}_1$ and $\mathbf{p}_2$, coupled to the electromagnetic field, described by the transverse vector potential $\mathbf{A}$. The corresponding Hamiltonian reads [41]

$$H(t) = \frac{\mathbf{p}_1^2}{2m_1} + \frac{\mathbf{p}_2^2}{2m_2} + V(|\mathbf{r}_1 - \mathbf{r}_2|) + H_{\text{int}}(t). \tag{1.106}$$

The time-independent part of this Hamiltonian describes the nuclear and Coulomb interaction between the two nuclei in the 'free' case (absence of the laser field). While the Hamiltonian describing the interaction of a single charge to a light field was already given in equation (1.49), in the case of two charges the interaction to the laser field can be reframed in the form

$$H_{\text{int}}(t) = -\sum_{i=1,2} \frac{Z_i e}{m_i}\mathbf{p}_i \cdot \mathbf{A}(\mathbf{r}_i, t) + \sum_{i=1,2} \frac{(Z_i e)^2}{2m_i}\mathbf{A}^2(\mathbf{r}_i, t). \tag{1.107}$$

By introducing a new set of variables, ($\mathbf{R}$, $\mathbf{r}$) for the coordinates, and ($\mathbf{P}$, $\mathbf{p}$) for the momenta, through the set of relations

$$\mathbf{r}_1 = \mathbf{R} + \frac{m_2}{m_1 + m_2}\mathbf{r}, \quad \mathbf{r}_2 = \mathbf{R} - \frac{m_1}{m_1 + m_2}\mathbf{r}, \tag{1.108}$$

$$\mathbf{p}_1 = \mathbf{p} + \frac{m_1}{m_1 + m_2}\mathbf{P}, \quad \mathbf{p}_2 = -\mathbf{p} + \frac{m_2}{m_1 + m_2}\mathbf{P}, \tag{1.109}$$

the transformation from the laboratory to the center-of-mass frame is accomplished.

If the two nuclei are assumed to form initially a quasi-bound state confined in a domain of size $R_{\text{N}} \sim r_0(A_1^{1/3} + A_2^{1/3})$, small compared to the radiation wavelength $\lambda = 2\pi c/\omega$ ($kR_{\text{N}} \ll 1$), the problem can be treated, as was shown earlier, in the dipole approximation. Since the vector potential is spatially homogeneous in this approximation,

$$\mathbf{A}(\mathbf{r}_1) \approx \mathbf{A}(\mathbf{r}_2) \approx \mathbf{A}(0), \tag{1.110}$$

and the interaction Hamiltonian is rewritten in the CM coordinates as

$$H_{\text{int}}(t) = -\frac{e}{m}\left[\frac{Z}{A}\mathbf{P} + \left(\frac{Z_1}{A_1} - \frac{Z_2}{A_2}\right)\mathbf{p}\right] \cdot \mathbf{A} + \frac{e^2}{2m}\left(\frac{Z_1^2}{A_1} + \frac{Z_2^2}{A_2}\right)\mathbf{A}^2, \tag{1.111}$$

where e is the proton charge and m is the nucleon mass.

Substituting the transformations (1.108)–(1.109) into (1.105), the time-dependent Schrödinger equation (TDSE) reads

$$\mathrm{i}\hbar\frac{\partial\psi(\mathbf{r}, \mathbf{R}, t)}{\partial t} = \left[\frac{\mathbf{p}^2}{2\mu} + \frac{\mathbf{P}^2}{2M} + V(\mathbf{r}) + H_{\text{int}}\right]\psi(\mathbf{r}, \mathbf{R}, t). \tag{1.112}$$

This equation can be further simplified by applying the *length gauge* which is appropriate in this case. The wavefunction is transformed according to

$$\psi(\mathbf{r}, \mathbf{R}, t) = \hat{U}_{\text{r}}\hat{U}_{\text{R}}\psi^{\text{L}}(\mathbf{r}, \mathbf{R}, t), \tag{1.113}$$

where

$$\hat{U}_{\mathrm{r}} = \exp\left[\frac{\mathrm{i}}{\hbar}eZ_{\mathrm{eff}}\mathbf{A}\cdot\mathbf{r}\right],\quad \hat{U}_{\mathrm{R}} = \exp\left[\frac{\mathrm{i}}{\hbar}eZ\mathbf{A}\cdot\mathbf{R}\right]. \tag{1.114}$$

Replacing the vector potential with the transverse electric field via

$$\mathbf{E}(t) = -\frac{\partial\mathbf{A}(t)}{\partial t}, \tag{1.115}$$

and introducing the notations $Z_{\mathrm{eff}} = (Z_1A_2 - Z_2A_1)/(A_1 + A_2)$ for the effective charge, and $\mu = m_1m_2/(m_1 + m_2)$ for the reduced mass of the dinuclear system, the two-body Schrödinger equation is reframed as

$$\begin{aligned} \mathrm{i}\hbar\frac{\partial\psi^{\mathrm{L}}(\mathbf{r},\mathbf{R},t)}{\partial t} = \Bigg[&-\frac{\hbar^2}{2\mu}\nabla_{\mathrm{r}}^2 - eZ_{\mathrm{eff}}\mathbf{r}\cdot\mathbf{E}(t) + V(\mathbf{r}) \\ &-\frac{\hbar^2}{2M}\nabla_{\mathrm{R}}^2 - eZ\mathbf{R}\cdot\mathbf{E}(t)\Bigg]\psi^{\mathrm{L}}(\mathbf{r},\mathbf{R},t). \end{aligned} \tag{1.116}$$

It can be observed that in the *length gauge* the interaction Hamiltonian couples the relative $\mathbf{r}$ and the center-of-mass $\mathbf{R}$ coordinates to the electric field $\mathbf{E}(t)$. The separation of these coordinates in (1.116) is accomplished through the factorization of the wavefunction

$$\psi^{\mathrm{L}}(\mathbf{r},\mathbf{R},t) = \phi^{\mathrm{L}}(\mathbf{r},t)\chi^{\mathrm{L}}(\mathbf{R},t). \tag{1.117}$$

Consequently, the time-dependent parabolic equation in the center-of-mass coordinate reads

$$\mathrm{i}\hbar\frac{\partial\chi^{\mathrm{L}}(\mathbf{R},t)}{\partial t} = \left[-\frac{\hbar^2}{2M}\nabla_{\mathrm{R}}^2 - eZ\mathbf{R}\cdot\mathbf{E}(t)\right]\chi^{\mathrm{L}}(\mathbf{R},t), \tag{1.118}$$

whereas in the relative coordinate

$$\mathrm{i}\hbar\frac{\partial\phi^{\mathrm{L}}(\mathbf{r},t)}{\partial t} = \left[-\frac{\hbar^2}{2\mu}\nabla_{\mathrm{r}}^2 + V(\mathbf{r}) - eZ_{\mathrm{eff}}\mathbf{r}\cdot\mathbf{E}(t)\right]\phi^{\mathrm{L}}(\mathbf{r},t). \tag{1.119}$$

The center-of-mass quantum dynamics of the dinuclear system in the presence of a radiation field, $\mathbf{A}(t)$, is described by Volkov states (see for example [42] for a comprehensive exposition). In order to make the problem easily tractable by avoiding the account of additional degrees of freedom, important for non-axial configurations ($\mathbf{E}\nparallel\mathbf{R}$), it is assumed that the expectation of $R(t)$ describes a trajectory collinear with the direction of the laser field. Thus, the R-dependent wavefunction χ is fully decoupled from the r-dependent decay wavefunction ϕ. In the case of a deformed daughter nucleus and/or a laser pulse with arbitrary polarization this simplification is less justified.

An alternative representation of the problem is provided by the application of the *velocity gauge* which consists in removing the term in $\mathbf{A}^2$ by means of the unitary transformation

$$\hat{U}_{\mathrm{V}} = \exp\left[\frac{\mathrm{i}}{\hbar}\frac{e^2}{2m}\left(\frac{Z_1^2}{A_1} + \frac{Z_2^2}{A_2}\right)\int \mathrm{d}t'\mathbf{A}^2(t')\right]. \tag{1.120}$$

The Schrödinger equation in the relative coordinate is represented in the form

$$\mathrm{i}\hbar\frac{\partial\phi^{\mathrm{V}}(\mathbf{r}, t)}{\partial t} = \left[-\frac{\hbar^2}{2\mu}\nabla_r^2 + V(\mathbf{r}) - \frac{1}{\mu}eZ_{\mathrm{eff}}\mathbf{p}\cdot\mathbf{A}(t)\right]\phi^{\mathrm{V}}(\mathbf{r}, t). \tag{1.121}$$

This Hamiltonian used above extends (1.82) to the two-body case.

Performing a time-dependent unitary transformation on top of the above *velocity gauge* wavefunction removes the $\mathbf{p}\cdot\mathbf{A}$ term in the interacting part of the Hamiltonian [38]. The relative motion wavefunction in the new representation (Kramers–Henneberger) is related to the original one by means of

$$\phi^{\mathrm{V}}(\mathbf{r}, t) = \exp\left[-\frac{\mathrm{i}}{\hbar}\vec{a}(t)\cdot\mathbf{p}\right]\phi^{\mathrm{K-H}}(\mathbf{r}, t), \tag{1.122}$$

where

$$\vec{a}(t) = -\frac{eZ_{\mathrm{eff}}}{\mu}\int \mathrm{d}t'\mathbf{A}(t') \tag{1.123}$$

is the vector corresponding to the displacement of the α-cluster from its oscillation center in the electric field $\mathbf{E}(t)$. In atomic physics this quantity is also known under the name of *quiver* motion.

The TDSE in the oscillating Kramers–Henneberger frame then reads

$$\mathrm{i}\hbar\frac{\partial\phi^{\mathrm{K-H}}(\mathbf{r}, t)}{\partial t} = \left[-\frac{\hbar^2}{2\mu}\nabla_{\mathrm{r}}^2 + V(\mathbf{r} + \vec{a}(t))\right]\phi^{\mathrm{K-H}}(\mathbf{r}, t). \tag{1.124}$$

The above form shows that in the K–H representation the effect of the external field is completely transferred into the argument of the static nuclear + Coulomb potential. Equation (1.124) provides the dynamics of the α-cluster in a moving frame of reference which follows the quiver motion $\vec{a}(t)$. In the K–H frame of reference the daughter nucleus experiences a quiver motion in the opposite direction, i.e. $-\vec{a}(t)$ [43].

1.7 Characterization of laser–nucleus interaction in analogy to atom–nucleus interaction

This book deals mainly with direct non-resonant laser–nucleus reactions [44]. Some examples are mentioned at the beginning of this chapter. It would be of certain practicality to review some of the laser parameters commonly used in the analysis of atomic effects induced by laser fields [45] and next perform a diagnosis of laser–nucleus reactions with the help of these parameters. To accomplish this task, instead of atomic units, nuclear adapted units for energy, time, size, etc, are introduced below.

The relation between the electric field strength $\mathscr{E}_0$ and the laser intensity $I_0 = c\varepsilon_0\mathscr{E}_0^2/2$ can be reframed for practical purposes as

$$\mathscr{E}_0\ [\mathrm{V\ cm^{-1}}] = 27.44\{I_0\ [\mathrm{W\ cm^{-2}}]\}^{1/2}. \tag{1.125}$$

It can be easily inferred from this last formula that at the maximum intensity foreseen at ELI [4], i.e. $I_0 \approx 10^{23}$ W cm^{-2}, the electric field is $\mathscr{E}_0 \approx 8.7 \times 10^{13}$ V cm^{-1} or, in nuclear units, $e\mathscr{E}_0 \approx 8.7 \times 10^{-6}$ MeV fm^{-1}. In this book we explore a wide range of laser intensities, including values above the ELI maximum.

As for the circular frequency, a convenient choice in nuclear units is [46]

$$\hbar\omega = 1240\ [\mathrm{MeV\ fm}]/\lambda\ [\mathrm{fm}]. \tag{1.126}$$

For an x-ray laser with wavelength of $\lambda = 1.24$ Å, the photon energy is $\hbar\omega = 10$ keV and the period $T = 2\pi/\omega = 4.13 \times 10^{-19}$ s $= 0.413$ attoseconds (as). In the course of the next chapters, laser photons with wavelengths in the range of XFEL and even smaller will prove to be good choices in order to explore the nuclear response. The intensities considered in this book are around or beyond the realm of values available at ELI. Since the main goal is to substantiate the magnification effect engendered by a superstrong laser pulse on a stable or radioactive nucleus, parameters for a fictitious laser facility will be assumed, particularly for cases involving radioactive decay and giant resonances. The production of such ultra-intense pulses is not far beyond the technical possibilities foreseen in the near future. For example pulses as short as 0.1 fs of x-rays of wavelength 1 Å, focused into a 1×1 nm^2 spot, and comprising 10^{12} photons (ten times larger than at SwissFEL), generate an energy release of 10^{-2} J, a power peak of 0.1 PW, and an intensity of $I_0 = 10^{28}$ W cm^{-2}.

Let us consider as an example the response of an α-radioactive nucleus to an incident laser field. As discussed earlier, the interaction is dominated by the electric dipole term: $-Z_{\mathrm{eff}}e\mathbf{E}(t)\cdot\mathbf{r}$. In close analogy to the atom–laser interaction case [45], the maximum value of the *ponderomotive energy* is defined as

$$U_{\mathrm{p}} = (1 + \xi^2)\frac{(Z_{\mathrm{eff}}e\mathscr{E}_0)^2}{4\mu\omega^2}. \tag{1.127}$$

It corresponds to the averaged kinetic energy gained by a free particle of mass μ and effective charge Z_{eff} once forced into oscillations by the laser field. For an intensity of 10^{29} W cm^{-2} and $\lambda = 1.24$ Å applied to the α-radioactive nucleus ^{106}Te, $U_{\mathrm{p}} \approx 2.9 \times 10^{-5}$ MeV. When $U_{\mathrm{p}} \approx \mu c^2$, relativistic effects start to be important. In equation (1.127) the ellipticity parameter ξ is 0 for a linearly polarized electromagnetic wave whereas $\xi = \pm 1$ for a circular one.

The square of the dimensionless parameter

$$\eta = \frac{v}{c} = \frac{eZ_{\mathrm{eff}}\mathscr{E}_0}{\mu\omega c} \tag{1.128}$$

represents the ratio of the maximum quiver velocity v of the laser-assisted nuclear system to the velocity of light; it indicates the onset of relativistic effects, when $\eta^2 \gtrsim 1$. For the XFEL-like parameters mentioned above, the non-relativistic regime is still effective since $\eta \approx 1.8 \times 10^{-4}$. On the other hand for the Ti:sapphire-like laser parameters, the wavelength is almost four orders of magnitude larger and for the same intensity the ponderomotive potential $U_p \approx 1200$ MeV approaches μc^2 and consequently the overunitary value of $\eta = 1.16$ signals the onset of relativistic effects.

Another example of a spontaneously decaying system is the spherical proton emitter system $^{171}\mathrm{Au} \rightarrow {}^{170}\mathrm{Pt} + p$. Following the study of Åberg [47], the nuclear radius assumes the value $R_N = 6.49$ fm, which reproduces the experimental proton decay energy $E_p = 1718 \pm 6$ keV (for a comprehensive discussion on the energy requirements for proton radioactivity see [48]). The effective charge is $Z_{\mathrm{eff}} = 0.538$. For hard x-ray photons with $\lambda = 0.124$ Å and a laser beam of intensity $I = 1.32 \times 10^{21}$ W cm^{-2} this choice corresponds to an extremely small ponderomotive energy, i.e. $U_p \approx 3 \times 10^{-8}$ eV. On the other hand, for the Ti:sapphire laser, $\lambda = 800$ nm ($\hbar\omega \approx 1.5$ eV), $I = 1.32 \times 10^{27}$ W cm^{-2}, $U_p \approx 126$ MeV, and $\eta \approx 0.71$. Thus, both cases are within the non-relativistic domain and the non-relativistic approach to the problem is justified. Like in the case of laser-assisted α-decay, relativistic effects are expected to be important once $U_p \geqslant \mu c^2 \approx 1$ GeV.

In his seminal paper [49] Keldysh argued that in a sufficiently strong low-frequency laser field the process of ionization is described by a tunneling formula adapted for a static electric field. For moments t close to $t_n = 2\pi n/\omega$ $(n = 1, 2, \ldots)$, such that $\cos\omega t \approx 1$, the total instantaneous potential of a quasi-bound proton perturbed by a laser pulse, with electric field oriented along the x-axis, is approximately $V(x) - eZ_{\mathrm{eff}}x\mathcal{E}_0$, and can be viewed as a static potential barrier. It is then justified to introduce the analogous of the atomic ionization energy I_p. In the case of a radioactive nucleus it can be equated to the difference between the height of the spherical Coulomb barrier and the proton decay energy, i.e. $I_p = V_{\max} - E_p$. For the above example I_p is close to 9 MeV. The same reasoning can be straightforwardly applied to any other type of emitted particle (n, d, α, etc). From I_p ensues another parameter of interest, the dimensionless Keldysh adiabaticity parameter. It was introduced by Keldysh in the context of the tunneling theory when $\hbar\omega \ll I_p$ [49]. Under such circumstances a Coulomb barrier is seen by a bound electron due to the *quasi-static* laser field acting on the atom. In the case of a proton filling a quasi-bound state inside the nucleus it is possible to lower the Coulomb barrier by increasing the strength of the external field. Following Keldysh, the barrier width in the nuclear context could be defined as

$$\Delta R = R_{t_2} - R_{t_1} \approx \frac{I_p}{Z_{\mathrm{eff}}e\mathcal{E}_0}, \tag{1.129}$$

where $R_{t_{1,2}}$ are the turning points of the Coulomb barrier. In the framework of the Keldysh approach, it is assumed that the sub-barrier motion of the nuclear particle is approximately the same with the motion of a free classical particle endowed with a velocity taken in average, i.e. $\frac{1}{2}\langle T\rangle = I_p$, to be

$$v_p \approx \sqrt{\frac{2I_p}{\mu}}, \tag{1.130}$$

such that the tunneling time is approximately equated to

$$t_{\text{tun}} \approx \frac{\Delta R}{v_{\text{p}}} \approx \sqrt{\frac{2\mu I_{\text{p}}}{Z_{\text{eff}} e \mathscr{E}_0}}. \tag{1.131}$$

If $t_{\text{tun}} < T$, the sign of the electric field does not change its value or sign during the tunneling time. Under such circumstances the tunneling takes place in a quasi-statical regime. The control parameter ωt_{tun} is dubbed the adiabaticity or Keldysh parameter,

$$\gamma_{\text{K}} = \omega \sqrt{\frac{2\mu I_{\text{p}}}{Z_{\text{eff}} e \mathscr{E}_0}}. \tag{1.132}$$

For low ω or large electric field strength $\mathscr{E}_0$, γ_K is small and the quasi-static or *optical tunneling* regime invoked by Keldysh is valid. In the case of atomic ionization it is commonly accepted that $\gamma \lesssim 1$ indicates that the tunneling process is fast on the atomic scale (*tunneling ionization*), whereas $\gamma \gtrsim 1$ corresponds to a slow tunneling process when multiphoton effects dominate (*multiphoton ionization*). Since in the case of a hard x-ray photon and lower intensity, $\gamma \approx 1.3 \times 10^7$, $U_{\text{p}} \ll \mu c^2$, the proton can leave the mother nucleus only via multiple absorption of photons. In this case the interaction between the dinuclear system and the laser field can be treated as a perturbation. On the other hand, at high laser intensities and for optical photons, $\gamma \approx 0.2$, the proton emission is expected to take place via tunneling through a modified Coulomb barrier determined by a large static electric field $\mathscr{E}_0$, such that V_{max} is lowered to $V_{\text{max}} - eZ_{\text{eff}}\mathscr{E}_0 R_b$ in the direction of the field at the barrier's peak position $r = R_b$. In this case a non-perturbative approach is adequate.

1.8 Doppler amplification of the laser pulse characteristics

As mentioned at the beginning of this chapter, state-of-the-art laser pulses from the optical domain are too low in frequency to make possible the direct access to nuclear states.

Another way to overcome the discrepancy between the photon energy of the available high-intensity lasers (<1 keV) and the energy scale of nuclear processes such as nuclear reactions, single-particle or collective excitations (~100 keV–20 MeV), was proposed in [50]. These authors put forward the proposal to accelerate heavy ions at ultra-relativistic energies in the presence of a super intense counter-propagating laser pulse. In this respect it is worthwhile to mention that the Large Hadron Collider (LHC), the world's largest and most powerful particle accelerator, has restarted and will be able to deliver for collisions, proton beams with the record energy of 13.6 trillion electronvolts (13.6 TeV) [51].

For a ion of charge state $Z - 1$ and of mass m_{ion} the Lorentz factor $\gamma = (1 - \beta^2)^{-1/2}$ is given by the formula [50]

$$\gamma = 1 + \frac{Z - 1}{m_{\text{ion}} c^2} E_{\text{acc}}. \tag{1.133}$$

In the case of counter-propagation of the laser pulse with respect to the heavy-ion beam, the observer moving with the former physical object will record at relativistic values of the scaled velocity $\beta = v/c$ an amplification of the electric and magnetic amplitudes of the electromagnetic field (**E**, **B**) [52]

$$\mathscr{E}' = \gamma(\mathscr{E} - v\mathscr{B}), \quad \mathscr{B}' = \gamma\left(\mathscr{B} - \frac{v}{c^2}\mathscr{E}\right), \tag{1.134}$$

and a Doppler shifted frequency

$$\omega' = \gamma(1 + \beta)\omega, \tag{1.135}$$

where the prime specifies the quantities in the moving frame. Assume that propagation takes along the x-axis, the electric field is oriented along the y-axis, whereas the magnetic induction is oriented along the z-axis. Since for a monochromatic plane wave $|\mathbf{E}|/|\mathbf{B}| = c$, the electromagnetic field amplitudes are transformed in the same way as the frequency, and since the 'peak intensity' $I(\omega)$ corresponding to a monochromatic plane wave of angular frequency ω is

$$I(\omega) = \frac{1}{2}c\varepsilon_0[\mathscr{E}(\omega)]^2, \tag{1.136}$$

its amplification, as seen by an observer fixed to the flying ion, is

$$I' = \frac{1 + \beta}{1 - \beta}I. \tag{1.137}$$

If one chooses the Ti:sapphire laser with photon energy $\hbar\omega = 1.5$ eV, propagating against a ^{208}Pb beam of ultra-relativistic energy $E_{\mathrm{acc}} = 10$ TeV, making thus $\gamma \approx 4182$ according to equation (1.133), the Lorentz transformation leads to the Doppler shifted energy quanta $\hbar\omega' \approx 12.54$ keV, and the large amplification of the intensity in the ion-fixed frame $I' \approx 7 \times 10^7 I$. If on the other hand a beam of hard x-rays with photon energy $\hbar\omega = 9.9$ keV and intensity $I = 10^{20}$ W cm^{-2} collides with the same ion beam, this time of energy $E_{\mathrm{acc}} = 1$ TeV ($\gamma \approx 419$), we end-up with $\hbar\omega' \approx 8.3$ MeV and $I' \approx 7 \times 10^{25}$ W cm^{-2}. A photon frequency shifted in the γ-ray range and an ultra-intense electric strength renders the excitation of nuclear collective levels by means of such a laser pulse feasible.

It is worthwhile to note than upon collision the ion beam emerges with scaled velocity

$$\beta' = \left(\beta - \frac{\hbar\omega}{m_{\mathrm{ion}}c^2}\right)\left(1 - \frac{\hbar\omega}{m_{\mathrm{ion}}c^2}\right)^{-1}. \tag{1.138}$$

It can be easily inferred that even in the unlikely circumstance of a nucleus absorbing $\approx$100 hard x-ray photons, the beam preserves its kinematical properties before the collision, i.e. $\beta' \approx \beta$.

References

[1] Strickland D and Mourou G A 1985 Compression of amplified chirped optical pulses *Opt. Commun.* **85** 219–21
[2] Mourou G A, Tajima T and Bulanov S V 2006 Optics in the relativistic regime *Rev. Mod. Phys.* **78** 309–71
[3] Announcement of the Nobel Prize in Physics 2018 https://www.youtube.com/user/thenobelprize/videos
[4] Extreme Light Infrastructure (ELI) http://www.eli-laser.eu
[5] Negoiţa F *et al* 2016 Laser driven nuclear physics at ELI-NP *Rom. Rep. Phys.* **68** S37–144
[6] Gales S 2016 The extreme light infrastructure-nuclear physics (ELI-NP) facility new horizons in physics with 10 PW ultra-intense lasers and 20 MeV brilliant gamma beams *Rep. Prog. Phys.* **81** 094301
[7] Schmüsser P, Dohlus M and Rossbach J 2008 *Ultraviolet and Soft X-ray Free-Electron Lasers* (Berlin: Springer)
[8] European XFEl 2011 http://xfel.eu
[9] Ishikawa T *et al* 2012 A compact x-ray free-electron laser emitting in the sub-Ångström region *Nat. Photon.* **6** 540–4
[10] Froideval A *et al* 2011 Towards possible opportunities in nuclear materials science and technology at an x-ray free electron laser research facility *J. Nucl. Mater.* **416** 242–51
[11] Mimura H *et al* 2014 Generation of 10^{20} W cm^{-2} hard x-ray laser pulses with two-stage reflective focusing system *Nat. Commun.* **5** 3539
[12] Hartmann F X, Nold W D and Sharon Y Y 1991 Model of nuclear excitation by electronic motion in intense laser fields *Phys. Rev.* A **44** 3210–21
[13] Typel S and Leclercq-Willain C 1996 Nuclear excitation by laser-assisted electronic transitions *Phys. Rev.* A **53** 2547–61
[14] Shabaz A, Müller C, Bürvenich T J and Keitel C H 2009 Laser-induced nonresonant nuclear excitation in muonic atoms *Nucl. Phys.* A **821** 106–17
[15] Dadi A and Müller C 2012 Laser-assisted nuclear photoeffect *Phys. Rev.* C **85** 064604
[16] Castañeda Cortés H M 2012 Laser assisted α-decay *Dissertation* University of Heidelberg p 135
Castañeda Cortés H M, Müller C, Keitel C H and Palffy A 2013 Nuclear recollisions in laser-assisted α decay *Phys. Lett.* B **723** 401–5
[17] Mişicu Ş and Rizea M 2013 α-decay in ultra-intense laser fields *J. Phys. G: Nucl. Part. Phys.* **40** 095101
[18] Mişicu Ş and Rizea M 2016 Speeding of α decay in strong laser fields *Open Phys.* **14** 81–7
[19] Kopytin I V and Kornev A S 2014 Semiclassical formula for the alpha-decay rate in an electromagnetic field *Phys. At. Nucl.* **77** 53–9
[20] Kis D P and Szilvasi R 2018 Three-dimensional α-tunneling in intense laser fields *J. Phys. G: Nucl. Part. Phys.* **45** 045103
[21] Bai D, Deng D and Ren Z 2018 Charged particle emissions in high-frequency alternative electric fields *Nucl. Phys.* A **976** 23–32
[22] Mişicu Ş and Rizea M 2019 Laser-assisted proton radioactivity of spherical and deformed nuclei *J. Phys. G: Nucl. Part. Phys.* **46** 115106
[23] Cheng J-H and Yu T-P 2022 Systematic study of laser-assisted proton radioactivity from deformed nuclei *Phys. Rev.* C **105** 024312

[24] Barna I F and Varró S 2015 Laser-assisted proton collision on light nuclei at moderate energies *Laser Part. Beams* **33** 299–305
[25] Barna I F and Varró S 2016 Proton scattering on carbon nuclei in bichromatic laser field at moderate energies *Nucl. Instrum. Methods Phys.* B **369** 77–82
[26] Mişicu Ş and Carstoiu F 2018 Fraunhofer and refractive scattering of heavy ions in strong laser fields *Eur. Phys. J.* A **54** 90
[27] Mişicu Ş 2022 Refractive scattering of loosely bound nuclei in arbitrarily polarized laser fields *Phys. Rev.* C **106** 034612
[28] Wang X 2020 Substantially enhanced deuteron-triton fusion probabilities in intense low-frequency laser fields *Phys. Rev.* C **102** 011601(R)
[29] Bekx J, Lindsey M L, Glenzer S H and Schlesinger K-G 2022 Applicability of semiclassical methods for modeling laser-enhanced fusion rates in a realistic setting *Phys. Rev.* C **105** 054001
[30] Bekx J, Lindsey M L and Schlesinger K-G 2022 Effect of nuclear charge on laser-induced fusion enhancement in advanced fusion fuels *Phys. Rev.* C **106** 034003
[31] Mişicu Ş 2022 Giant dipole multi-resonances excited by high-frequency laser pulses *Particles* **5** 287–97
[32] Diels J-C and Rudolph W 2006 *Ultrashort Laser Pulse Phenomena* 2nd edn (Amsterdam: Elsevier)
[33] Jackson J D 1999 *Classical Electrodynamics* 3rd edn (Hoboken, NJ: Wiley)
[34] Goldstein H 1952 *Classical Mechanics* (Cambdridge, MA: Addison-Wesley)
[35] Reiss H R 2008 Foundations of the strong-field physics *Lectures on Ultrafast Intense Lase Science* ed K Yamanouchi, S L Chin, P Agostini and G Ferrante vol 3 (Berlin: Springer) ch 1 pp 1–31
[36] Reiss H R 2009 Dependence of physical interpretation on the choice of electromagnetic gauge *Laser Phys.* **19** 1626–34
[37] Reiss H R 2008 Limits on tunneling theories of strong-field ionization *Phys. Rev. Lett.* **101** 043002
[38] Henneberger W C 1968 Perturbation method for atoms in intense light beams *Phys. Rev. Lett.* **21** 838–41
[39] Kramers H A 1950 Non-relativistic quantum-electrodynamics and correspondence principle *Proc. Solvay Conf. 'Les Particules Elémentaires' (1948)* (Solvay Foundation) pp 241–65
[40] Wolkow D M 1935 Über eine Klasse von Lösungen der Diracschen Gleichung *Z. Phys.* **94** 250–60
[41] Cohen-Tannoudji C, Dupont-Roc J and Grynberg G 1997 *Photons and Atoms: Introduction to Quantum Electrodynamics* (New York: Wiley)
[42] Mittleman M 1993 *Introduction to the Theory of Laser–Atom Interactions* 2nd edn (New York: Plenum)
[43] Gavrila M 1992 Atomic structure and decay in high-frequency fields *Atoms in Intense Laser Fields* ed M Gavrila (Boston, MA: Academic) pp 435–510
[44] Di Piazza A, Müller C, Hatsagortsyan K Z and Keitel C H 2012 Extremely high-intensity laser interactions with fundamental quantum systems *Rev. Mod. Phys.* **84** 1177–228
[45] Mulser P and Bauer D 2010 *High Power Laser–Matter Interaction* (*Springer Tracts in Modern Physics*) vol 238 (Berlin: Springer)
[46] Habs D, Günther M M, Jentschel M and Thirolf P G 2012 Nuclear photonics *AIP Conf. Proc.* **1462** 177–84

[47] Åberg S, Semmes P B and Nazarewicz W 1997 Spherical proton emitters *Phys. Rev.* C **56** 1762–73
Erratum: 1997 *Phys. Rev.* C **56** 1762
[48] Hofmann S 1996 Proton radioactivity *Nuclear Decay Modes* ed D N Poenaru (Bristol: Institute of Physics Publishing) ch 3 pp 143–203
[49] Keldysh L V 1965 Ionization in the field of a strong electromagnetic wave *Sov. Phys. JETP* **20** 1307–14
[50] Mocken G R and Keitel C H 2004 Bound atomic dynamics in the MeV regime *J. Phys.* B **37** L275–83
[51] 2022 Large Hadron Collider restarts *CERN News* 22 Apr. https://home.cern/news/news/accelerators/large-hadron-collider-restarts
[52] Landau L D and Lifshitz E M 1962 *The Classical Theory of Fields* (Oxford: Pergamon)
Landau L D and Lifshitz E M 1989 *Théorie des Champs* (Moscow: Éd. Mir Moscou)

Chapter 2

Theoretical treatment of classical and quantum systems in strong laser fields

The aim of this chapter is to present some of the theoretical methods used to describe the coupling of classical and quantum systems to an external electromagnetic field. Although these methods were first developed to describe the interaction of plasmas and atoms with laser fields, the extension to the cases of nuclear structure and nuclear reactions can be done in a straightforward manner. A first example is provided by the oscillations induced by an external periodic electric field in a classical plasma and there is the scope to extend the analogy in the next chapter to the giant resonances within the hydrodynamical model of the atomic nucleus. The next part of this chapter presents the time-dependent perturbative and non-perturbative methods used to describe the interaction of quantum systems, such as atoms and nuclei, with external electromagnetic fields.

2.1 Oscillations of a classical plasma induced by a periodic electric field

Consider a classical multicomponent plasma with two constituents of number density ρ_q and electric charge e_q ($q = 1, 2$) in the unperturbed state. If a variable electric field is applied to the plasma in the form

$$\mathbf{E}(t) = \hat{\varepsilon}\mathscr{E}_0 \sin \omega t, \tag{2.1}$$

and the thermal motion of the charged particles is neglected, then, based on the assumption that the energy of oscillations generated by the external perturbation exceeds the thermal energy (assumption of a cold plasma), the dynamics can be described in the hydrodynamical framework for a fluid mixture of charged particles [1]:

$$\frac{\partial \rho_q}{\partial t} + \nabla \cdot (\rho_q \mathbf{v}_q) = 0, \tag{2.2}$$

doi:10.1088/978-0-7503-3199-9ch2

© IOP Publishing Ltd 2023

$$\frac{\partial \mathbf{v}_q}{\partial t} + (\mathbf{v}_q \cdot \nabla)\mathbf{v}_q = \frac{e_q}{m_q}\mathbf{E}. \tag{2.3}$$

The magnetic component of the Lorentz force $e_q(\mathbf{v}_q \times \mathbf{B})$, acting on the charge e_q, was neglected in the above formula. In the homogeneous state of the mixture the current is absent, i.e. $\nabla \cdot (\rho_q \mathbf{v}_q) = 0$, and the particles of type q move along the electric field lines with velocity

$$\mathbf{u}_q = -\frac{e_q}{m_q \omega}\mathbf{E}\cos \omega t. \tag{2.4}$$

In order to study oscillatory phenomena in an infinitely extended mixture the plane wave assumption is adopted,

$$\mathbf{u}_q \sim \mathrm{e}^{\mathrm{i}\mathbf{k}\cdot\mathbf{r}}, \tag{2.5}$$

and small departures of the dynamical variable from equilibrium values are allowed,

$$\rho_q'(t) = \rho_q + \delta\rho_q(t), \qquad \mathbf{v}_q'(t) = \delta\mathbf{v}_q(t), \tag{2.6}$$

where the second relation reflects the fact that in the equilibrium state the two species are at rest. Using the first Maxwell equation, relating the fluctuation of the electric field in the plasma with the sum of the density fluctuation components,

$$\nabla \cdot \delta\mathbf{E} = \frac{1}{\varepsilon_0}\sum_q e_q \delta\rho_q, \tag{2.7}$$

the following system of coupled second-order differential equations, describing the evolution of the particle number density fluctuations is obtained:

$$\left(\frac{\partial}{\partial t} + \mathrm{i}\mathbf{k}\cdot\mathbf{u}_q\right)^2 \delta\rho_q(t) = -\frac{e_q\rho_q}{m_q\varepsilon_0}\sum_{q'=1,2} e_{q'}\delta\rho'_q(t) \qquad (q = 1, 2). \tag{2.8}$$

Introducing in place of $\delta\rho_q$ the new functions

$$\nu_q(t) = e_q\delta\rho_q \mathrm{e}^{-\mathrm{i}\frac{e_q\mathscr{E}_0}{m_q\omega^2}\hat{\varepsilon}\cdot\mathbf{k}\sin\omega t}, \tag{2.9}$$

representing the number density of q-type particles in the coordinate system oscillating with the particles of that sort, Aliev and Silin [2] obtained a system of two coupled Hill equations [3, 4],

$$\ddot{\nu}_1 + \omega_1^2(\nu_1 + \nu_2\mathrm{e}^{-\mathrm{i}\alpha_0\hat{\varepsilon}\cdot\mathbf{k}\sin\omega t}) = 0, \tag{2.10}$$

$$\ddot{\nu}_2 + \omega_2^2(\nu_2 + \nu_1\mathrm{e}^{\mathrm{i}\alpha_0\hat{\varepsilon}\cdot\mathbf{k}\sin\omega t}) = 0, \tag{2.11}$$

where

$$\alpha_0 = -\left(\frac{q_1}{m_1} + \frac{q_2}{m_2}\right)\frac{\mathscr{E}_0}{\omega^2}, \tag{2.12}$$

and

$$\omega_q = \left(\frac{e_q^2 \rho_\alpha}{m_q \varepsilon_0} \right)^{1/2} \tag{2.13}$$

is the qth component *plasma frequency*.

Applying the Floquet technique, discussed below in this chapter, the solutions of the coupled system (2.10) and (2.11) are represented in the form

$$\nu_q(t) = \mathrm{e}^{-\mathrm{i}\Omega t} \sum_{n=-\infty}^{\infty} \rho_q^{(n)} \mathrm{e}^{-\mathrm{i}n\omega t}. \tag{2.14}$$

Next, making use of the Jacobi–Anger identity (A.31), a set of equations for the coefficients of the Fourier expansion (2.14) is obtained:

$$\rho_1^{(n)} - \frac{\omega_1^2}{(\Omega + n\omega)^2 - \omega_1^2} \sum_{n'=-\infty}^{\infty} J_{n-n'}(\alpha_0 \hat{\varepsilon} \cdot \mathbf{k}) \rho_2^{(n')} = 0, \tag{2.15}$$

$$\rho_2^{(n)} - \frac{\omega_2^2}{(\Omega + n\omega)^2 - \omega_2^2} \sum_{n'=-\infty}^{\infty} J_{n'-n}(\alpha_0 \hat{\varepsilon} \cdot \mathbf{k}) \rho_1^{(n')} = 0. \tag{2.16}$$

In the high-frequency case, i.e. $\omega \gg \omega_{1,2}$, the term with $n = 0$ dominates over the terms with $n \neq 0$ in the above sums. Thus

$$\rho_1^{(0)} = \frac{\omega_1^2}{\Omega^2 - \omega_1^2} J_0(\alpha_0 \hat{\varepsilon} \cdot \mathbf{k}) \rho_2^{(0)}, \tag{2.17}$$

$$\rho_2^{(0)} = \frac{\omega_2^2}{\Omega^2 - \omega_2^2} J_0(\alpha_0 \hat{\varepsilon} \cdot \mathbf{k}) \rho_1^{(0)}, \tag{2.18}$$

which yields directly the dispersion relation

$$\omega_1^2 \omega_2^2 J_0^2(\alpha_0) = (\Omega^2 - \omega_1^2)(\Omega^2 - \omega_2^2). \tag{2.19}$$

Consider the case of a rarefied gas of protons and neutrons. The effective charges of the two species are [5]

$$q_1 = \frac{eN}{A}, \qquad q_2 = -\frac{eZ}{A}. \tag{2.20}$$

Introducing the notation $Y = \rho_1/\rho_2$, the dispersion relation has two branches in this case:

$$\Omega_{1,2}^2 = \frac{1}{2}\omega_1^2 \left[1 + Y \pm \sqrt{(1 + Y)^2 - 4Y(1 - J_0^2(\alpha_0 \hat{\varepsilon} \cdot \mathbf{k}))} \right]. \tag{2.21}$$

For symmetric nuclear matter, $\rho_1 = \rho_2$, and the above roots are rewritten as

$$\Omega_{1,2}^2 = \omega_1^2(1 \pm J_0^2(\alpha_0\hat{\varepsilon}\cdot\mathbf{k})). \tag{2.22}$$

For small arguments of the Bessel function the ascending series formula (A.30) from the appendix yields

$$\Omega_1^2 \approx \omega_1^2\left(2 - \frac{1}{2}(\alpha_0\hat{\varepsilon}\cdot\mathbf{k})^2\right), \qquad \Omega_2^2 \approx \omega_1^2(\alpha_0\hat{\varepsilon}\cdot\mathbf{k})^2. \tag{2.23}$$

Thus, in the high-frequency approximation two branches of the dispersion relation are relevant. A similar case will be encountered in chapter 3, where a splitting of the giant dipole electric resonance takes place in a high-frequency electric continuous wave.

2.2 Perturbative methods

2.2.1 Nonstationary perturbation theory

The time-dependent theory of perturbations applies to a quantum system that at an initial moment t_i is perturbed by an external field H_{int} such that at a later time t it evolves in accordance with the Schrödinger equation

$$\mathrm{i}\hbar\frac{\partial\Psi(t)}{\partial t} = [H_0 + H_{\text{int}}(t)]\Psi(t). \tag{2.24}$$

The unperturbed (*field-free*) system is described by the Hamiltonian H_0 and its eigenvalues $E_n = \hbar\omega_n$, and the eigenfunctions $\psi_n^{(0)}$, including also states in the continuum, are given by the stationary Schrödinger equation

$$H_0\psi_n^{(0)} = E_n\psi_n^{(0)}. \tag{2.25}$$

Assuming that at $t = -\infty$ the system is described by the stationary state $\Psi(t = 0) = \psi_i^{(0)}$, the object of the perturbative calculation is to determine the probabilities that the system remains in the same state or evolves in another eigenstate $\psi_f^{(0)}$,

$$p(i \to f) = \left|\left\langle\psi_f^{(0)}\middle|\Psi(t)\right\rangle\right|^2. \tag{2.26}$$

Since this probability implicitly contains the information contained in the eigenstates $\psi_i^{(0)}$ and $\psi_f^{(0)}$, it is then reasonable to write the perturbed wavefunction as a superposition of wavefunctions from the field-free basis,

$$\Psi(t) = \sum\!\!\!\!\!\!\int_n C_n(t)\mathrm{e}^{-\mathrm{i}E_nt/\hbar}\psi_n^{(0)}, \tag{2.27}$$

where the sum over n includes the summation over bound states and integration over states in the continuum. The time-dependent transition (excitation) amplitudes are

$$C_n(t) = \left\langle\psi_n^{(0)}\middle|\Psi(t)\right\rangle\mathrm{e}^{\mathrm{i}E_nt/\hbar}. \tag{2.28}$$

By substituting equation (2.27) in equation (2.24),

$$\mathrm{i}\hbar\frac{\mathrm{d}}{\mathrm{d}t}\sum\!\!\!\!\!\!\!\int_n C_n(t)\mathrm{e}^{-\mathrm{i}E_nt/\hbar}\psi_n^{(0)} = [H_0 + H_{\mathrm{int}}(t)]\sum\!\!\!\!\!\!\!\int_n C_n(t)\mathrm{e}^{-\mathrm{i}E_nt/\hbar}\psi_n^{(0)}. \tag{2.29}$$

The left-hand side of this last equation can be rewritten with the help of the product rule for differentiation, as

$$\begin{aligned}&\mathrm{i}\hbar\sum\!\!\!\!\!\!\!\int_n\left[\dot{C}_n(t)\mathrm{e}^{-\mathrm{i}E_nt/\hbar}\psi_n^{(0)} + C_n(t)\frac{\mathrm{d}}{\mathrm{d}t}\left(\mathrm{e}^{-\mathrm{i}E_nt/\hbar}\psi_n^{(0)}\right)\right]\\ &= [H_0 + H_{\mathrm{int}}(t)]\sum\!\!\!\!\!\!\!\int_n C_n(t)\mathrm{e}^{-\mathrm{i}E_nt/\hbar}\psi_n^{(0)}.\end{aligned} \tag{2.30}$$

On the other hand, multiplying equation (2.25) by $C_n(t)\mathrm{e}^{-\mathrm{i}E_nt/\hbar}$ and summing/integrating over n, the following identity is obtained:

$$\mathrm{i}\hbar\sum\!\!\!\!\!\!\!\int_n C_n(t)\frac{\mathrm{d}}{\mathrm{d}t}\left(\mathrm{e}^{-\mathrm{i}E_nt/\hbar}\psi_n^{(0)}\right) = \sum\!\!\!\!\!\!\!\int_n C_n(t)\mathrm{e}^{-\mathrm{i}E_nt/\hbar}H_0\psi_n^{(0)}, \tag{2.31}$$

that, once inserted in equation (2.30), provides

$$\sum\!\!\!\!\!\!\!\int_n \dot{C}_n(t)\mathrm{e}^{-\mathrm{i}E_nt/\hbar}\psi_n^{(0)} = -\frac{i}{\hbar}\sum\!\!\!\!\!\!\!\int_n C_n(t)\mathrm{e}^{-\mathrm{i}E_nt/\hbar}H_{\mathrm{int}}(t)\psi_n^{(0)}. \tag{2.32}$$

Next, by multiplying both sides of this last equation with $\psi_m^{(0)*}$ and integrating over space we arrive at a system of linear differential equations allowing thus the determination of the unknown coefficients C_n,

$$\dot{C}_m(t) = -\frac{\mathrm{i}}{\hbar}\sum\!\!\!\!\!\!\!\int_n C_n(t)V_{mn}(t)\mathrm{e}^{\mathrm{i}(E_m-E_n)t/\hbar}, \tag{2.33}$$

where

$$V_{mn}(t) = \left\langle\psi_m^{(0)}\middle|H_{\mathrm{int}}(t)\middle|\psi_n^{(0)}\right\rangle \tag{2.34}$$

are the matrix elements of the interaction Hamiltonian. The above set of coupled equations is solved by assuming that at time $t_{\mathrm{i}} = 0$ (usually for bound states) or $t_{\mathrm{i}} = -\infty$ (usually for scattering problems), the nucleus is in its ground state, i.e.

$$C_n(t_{\mathrm{i}}) = \delta_{n0}. \tag{2.35}$$

The values of C_n at $t_f = +\infty$ are the excitation amplitudes after the perturbation H_{int} is switched off. Thus, the excitation probability (2.26) of the nth state is given by

$$p(0 \to n) = |C_n|^2. \tag{2.36}$$

It satisfies the conservation of total probability at any time

$$\sum_n p(0 \to n) = \sum_n |C_n|^2 = 1. \tag{2.37}$$

The approximation of the perturbation theory consists in writing the solution $C_n(t)$ in the form of a series in powers of the strength of the interaction

$$C_n = C_n^{(0)} + C_n^{(1)} + C_n^{(2)} + \cdots, \tag{2.38}$$

where $C_n^{(0)}$ are the coefficients in the zero-order approximation not depending on V. The coefficients in the first-order approximation $C_n^{(1)}$ depend on V, $C_n^{(2)}$ depend on V^2, and so on.

Substituting (2.38) in (2.33) and keeping only the terms up to first-order, i.e. $C_n = C_n^{(0)} + C_n^{(1)}$,

$$\dot{C}_n^{(0)} = 0, \tag{2.39}$$

$$\dot{C}_n^{(1)} = -\frac{\mathrm{i}}{\hbar}\sum_{n'} V_{nn'}(t)C_{n'}^{(0)}\mathrm{e}^{\mathrm{i}(E_n - E'^n)t/\hbar}. \tag{2.40}$$

If at $t = -\infty$ the quantum system is found in the eigenstate with label n, i.e. $C_n(t) = 1$ and $C_{m\neq n} = 0$, then the first of the above equations provides the solution

$$C_n^{(0)} = 1. \tag{2.41}$$

The second equation, (2.40), substituted in the truncated series (2.38), gives upon integration

$$C_n^{(1)}(t) = -\frac{\mathrm{i}}{\hbar}\int_{-\infty}^{t} \mathrm{d}t' \mathrm{e}^{\mathrm{i}(E_n - E'^n)t'/\hbar} V_{nn'}(t'). \tag{2.42}$$

The probability to observe a transition from the initial state n to a new state m after a time t is

$$p(n \to m) = |C_m(t)|^2. \tag{2.43}$$

2.2.2 Perturbations periodic in time

The application of the first-order time-dependent perturbation theory is illustrated for the case of a particle of mass m and charge q moving in a one-dimensional harmonic potential with frequency ω_0. Initially the particle is found in the ground state of the potential $V(x) = m\omega_0^2 x^2/2$ and at $t = -\infty$ a pulse of electric field $E(t) = \mathscr{E}_0 \mathrm{e}^{-t^2/\tau_p^2} \sin \omega t$, of frequency ω and duration τ_p, acts upon it. The interaction between the particle and the field is evidently $H_{\mathrm{int}} = -qxE(t)$. Since the particle occupies initially the ground state of the harmonic potential, at asymptotic time, when the electrical pulse has died out, the probability to populate excited states, i.e. depopulate the ground state, is given by a sum performed over all excited nuclear states:

$$P(t \to \infty) = \sum_{m\neq 0} p(0 \to m). \tag{2.44}$$

The transition matrix elements of the interaction is

$$\begin{aligned} V_{m0} &\equiv \langle m|H_{\text{int}}|0\rangle \\ &= q\mathscr{E}_0 e^{-t^2/\tau_p^2} \sin \omega t \langle m|x|0\rangle = \delta_{m,1} q\mathscr{E}_0 \left(\frac{\hbar}{2m\omega_0}\right)^{1/2} e^{-t^2/\tau_p^2} \sin \omega t. \end{aligned} \tag{2.45}$$

Combining equations (2.42), (2.43), and (2.45) provides the probability (2.44),

$$\begin{aligned} P(t \to \infty) &= \frac{1}{\hbar^2} \left| \int_{-\infty}^{+\infty} \mathrm{d}t V_{10}\, e^{i\omega_0 t} \right|^2 \\ &= \frac{\hbar}{2m\omega_0} \left(\frac{q\mathscr{E}_0}{\hbar}\right)^2 \left| \int_{-\infty}^{+\infty} \mathrm{d}t e^{-\frac{t^2}{\tau_p^2}}\, e^{i\omega_0 t} \sin \omega t \right|^2. \end{aligned} \tag{2.46}$$

Using the integral identity [6]

$$\int_{-\infty}^{+\infty} \mathrm{d}x\ e^{-a^2x^2 + ibx} = \frac{\sqrt{\pi}}{a} e^{-\frac{b^2}{4a^2}}, \tag{2.47}$$

equation (2.46) finally acquires the form

$$P(t \to \infty) = \left(\frac{\pi q^2 \mathscr{E}_0^2}{2m\hbar\omega_0}\right) \tau_p^2 e^{-\frac{1}{2}(\omega^2 + \omega_0^2)\tau_p^2} \sinh^2\!\left(\frac{1}{2}\omega\omega_0\tau_p^2\right). \tag{2.48}$$

For pulses of long duration

$$P(t \to \infty) \approx \left(\frac{\pi q^2 \mathscr{E}_0^2}{2m\hbar\omega_0}\right) \tau_p^2 e^{-\frac{1}{2}(\omega^2 - \omega_0^2)\tau_p^2}. \tag{2.49}$$

In this approximation the maximum of the transition probability to the first excited state is obtained when the value of τ_p is

$$\tau_p \approx \sqrt{\frac{2}{|\omega^2 - \omega_0^2|}}. \tag{2.50}$$

The corresponding value of the transition probability is obtained in the non-resonant case ($\omega \neq \omega_0$) by substituting equation (2.50) in equation (2.49),

$$P(t \to \infty) \approx \frac{\pi}{e} \frac{q^2 \mathscr{E}_0^2}{m\hbar\omega_0 |\omega^2 - \omega_0^2|}. \tag{2.51}$$

In the resonant case ($\omega = \omega_0$) two cases are valid.

When $\tau_p > \omega_0^{-1}$

$$P(t \to \infty) \approx \left(\frac{\pi q^2 \mathscr{E}_0^2}{8m\hbar\omega_0}\right) \tau_p^2, \tag{2.52}$$

whereas for $\tau_p < \omega_0^{-1}$

$$P(t \to \infty) \approx \left(\frac{\pi q^2 \mathscr{E}_0^2}{8m\hbar}\right) \omega_0 \tau_p^4. \tag{2.53}$$

2.2.3 Adiabatic perturbation

In the case of a nuclear quantum system, the optical or even a high-frequency x-ray laser acting as a perturbation varies slowly in time compared to the time-scale of a nuclear process. Therefore the system is adjusting adiabatically to the slow change of external parameters, i.e. the electric strength of the laser field in this case.

The solution of the Schrödinger equation in this case is sought in the form of a superposition of stationary eigenstates $\psi_m(x, \xi)$

$$\Psi = \sum_m C_m(t)\psi_m(x, \xi)e^{-\frac{i}{\hbar}\int_{-\infty}^{t} \varepsilon_m(\xi)dt}. \tag{2.54}$$

It corresponds to an arbitrary but fixed value of the parameter ξ that satisfies the eigenvalue problem

$$H(x, \xi)\psi_n(x, \xi) = \varepsilon_n(\xi)\psi_n(x, \xi). \tag{2.55}$$

Substituting equation (2.54) in the TDSE,

$$\frac{dC_n(t)}{dt} + \sum_m \left\langle \psi_n \middle| \frac{\partial \psi_m}{\partial \xi} \right\rangle \dot{\xi} C_m e^{-\frac{i}{\hbar}\int_{-\infty}^{t} (\varepsilon_m - \varepsilon_n)dt} = 0. \tag{2.56}$$

Since the perturbation is assumed to change adiabatically, then $\dot{\xi}$ is small and the above infinite system of first-order linear differential equations can be solved by successive approximations:

$$C_n^{(0)}(t) = C_n(-\infty), \tag{2.57}$$

$$C_n^{(1)}(t) = -\int_{-\infty}^{t} dt' \sum_m \left\langle \psi_n \middle| \frac{\partial \psi_m}{\partial \xi} \right\rangle \dot{\xi} C_m(-\infty) e^{-\frac{i}{\hbar}\int_{-\infty}^{t'} (\varepsilon_m - \varepsilon_n)dt}. \tag{2.58}$$

Differentiating (2.55) with respect to ξ, multiplying the obtained equation from the left by ψ_n and integrating over x, the overlap coefficient $\langle \psi_n | \partial \psi_m / \partial \xi \rangle$ is expressed in terms of the matrix element of the Hamiltonian H,

$$\left\langle \psi_n \middle| \frac{\partial \psi_m}{\partial \xi} \right\rangle = -\frac{\left\langle \psi_n \middle| \frac{\partial H}{\partial \xi} \middle| \psi_m \right\rangle}{\varepsilon_m - \varepsilon_n}, \qquad n \neq m. \tag{2.59}$$

The wavefunction ψ_n is assumed to be real and non-degenerate, and the diagonal element vanishes, i.e.

$$\left\langle \psi_n \middle| \frac{\partial \psi_n}{\partial \xi} \right\rangle = \frac{1}{2}\frac{\partial}{\partial \xi} \int \psi_n^2 dx = 0. \tag{2.60}$$

Finally, the expansion coefficients in the first-order approximation are represented in the form

$$C_n^{(1)}(t) = -\int_{-\infty}^{t} \mathrm{d}t' \sum_{m \neq n} \frac{\left\langle \psi_n \left| \frac{\partial H}{\partial \xi} \right| \psi_m \right\rangle}{\varepsilon_m - \varepsilon_n} \dot{\xi} C_m(-\infty) \mathrm{e}^{-\frac{\mathrm{i}}{\hbar}\int_{-\infty}^{t'} (\varepsilon_m - \varepsilon_n)\mathrm{d}t}. \tag{2.61}$$

2.3 Non-perturbative methods

2.3.1 Exact treatment of the harmonic oscillator in a pulsed electromagnetic field

Consider a harmonic oscillator linearly coupled to a periodic or almost periodic external field [7], described by the time-dependent Schrödinger equation

$$\mathrm{i}\hbar \frac{\partial \psi(\xi, t)}{\partial t} = -\frac{\hbar^2}{2B}\frac{\partial^2 \psi(\xi, t)}{\partial \xi^2} + \left[\frac{1}{2}B\omega_0^2\xi^2 - \xi f(t)\right]\psi(\xi, t). \tag{2.62}$$

Such a time-dependent quantum model can be applied to the generation of dipole oscillations of the electron cloud–nucleus system induced by the electric component of a linearly polarized laser pulse acting in the direction specified by the unit vector $\mathbf{e}_\xi$ [8],

$$\mathbf{E}(t) = \mathbf{e}_\xi \mathscr{E}_0 F(t) \cos \omega t, \tag{2.63}$$

where $F(t)$ is the shape (envelope) of the pulse. Two envelopes, already introduced in the previous chapter, are employed in what follows. Both start at $t = 0$ and posses the following profiles:

$$\text{(I)}\;\; F(t) = \Theta(\tau_p - t), \quad \text{(II)}\;\; F(t) = \sin^2\left(\frac{\pi t}{\tau_p}\right)\Theta(\tau_p - t), \tag{2.64}$$

where τ_p is the pulse duration which can be an integer multiple of the laser period

$$\tau_p = nT. \tag{2.65}$$

The functional form (I), known as the *continuous wave* (CW) is easy to handle, while the more realistic form (II) can also be easily treated since a trivial Fourier decomposition of the signal (2.63) in a sum of three modes of frequencies ω, $\omega - 2\Omega$, $\omega + 2\Omega$ with $\Omega = 2\pi/\tau_p$, is allowed,

$$F(t)\cos \omega t = \frac{1}{2}\left[\cos \omega t - \frac{1}{2}\cos(\omega - 2\Omega)t - \frac{1}{2}\cos(\omega + 2\Omega)t\right]\Theta(\tau_p - t). \tag{2.66}$$

The solution of equation (2.62) at time t can be obtained easily via the general method of Husimi [9] for an oscillation frequency variable in time and by Kerner [10] for constant oscillation frequency. More details can be found in [7]. A solution of (2.62) can be designed such that it is in a one-to-one correspondence with the free stationary solution Φ, when $f(t) = 0$:

$$\psi(\xi, t) = \Phi(\xi - u(t))e^{\frac{i}{\hbar}[B\dot{u}(t)(\xi - u(t)) + \mathcal{S}(t)]}. \tag{2.67}$$

The unknown functions $u(t)$ and $\mathcal{S}(t)$ are determined by substitution of the Ansatz (2.67) in equation (2.62), and by requiring that Φ satisfies the time-dependent Schrödinger equation for the unperturbed harmonic oscillator of shifted coordinate, $\xi - u(t)$,

$$i\hbar\frac{\partial\Phi}{\partial t} = -\frac{\hbar^2}{2B}\frac{\partial^2\Phi}{\partial(\xi - u(t))^2} + \frac{1}{2}B\omega_0^2(\xi - u(t))^2\Phi. \tag{2.68}$$

The scalar time-dependent function $u(t)$ obeys the classical equation of motion for the one-dimensional forced harmonic oscillator

$$\ddot{u}(t) + \omega_0^2 u(t) = \frac{1}{B}f(t) \tag{2.69}$$

and

$$\mathcal{S}(t) = \int_{-\infty}^{t} \mathcal{L}(t')\mathrm{d}t', \tag{2.70}$$

is precisely the action integral corresponding to the classical Lagrangian of the driven oscillator

$$\mathcal{L}(t) = \frac{1}{2}B\dot{u}^2(t) - \frac{1}{2}B\omega_0^2 u^2(t) + f(t)u(t). \tag{2.71}$$

For the rectangular envelope

$$f(t) = q\mathcal{E}_0 \cos \omega t \Theta(\tau_p - t), \tag{2.72}$$

the solution of equation (2.69), with starting values $u(0) = 0$ and $\dot{u}(0) = 0$, is [11]

$$u(t) = \frac{\alpha_0\omega^2}{\omega^2 - \omega_0^2}(\cos \omega_0 t - \cos \omega t), \tag{2.73}$$

where

$$\alpha_0 = \frac{q\mathcal{E}_0}{B\omega^2} \tag{2.74}$$

is the analogous of the quiver amplitude of the free electron in a light field [12]. In this case it represents the maximum displacement of the oscillator center due to the action of the external periodic electric field.

For $t > \tau_p$, the external force ceases to act on the oscillator and the displacement field is

$$u(t) = A \sin \omega_0 t + B \cos \omega_0 t, \tag{2.75}$$

where A and B are determined from the matching continuity of u and its first-order derivative at $t = \tau_p$.

At resonance ($\omega = \omega_0$), the displacement field (2.73) reads

$$\lim_{\omega\to\omega_0} u(t) = \frac{q\mathscr{E}_0}{2B\omega_0} t \sin\omega_0 t. \tag{2.76}$$

For the '$\sin^2$' envelope,

$$f(t) = \frac{1}{2} q\mathscr{E}_0 \left[\cos\omega t - \frac{1}{2}\cos\omega_+ t - \frac{1}{2}\cos\omega_- t\right]\Theta(\tau_p - t), \tag{2.77}$$

and introducing the notation $\omega_\pm = \omega \pm 2\Omega$, the solution of equation (2.69) is

$$\begin{aligned} u(t) = &\frac{1}{2}\alpha_0\omega^2 \\ &\left[\frac{\cos\omega t - \cos\omega_0 t}{\omega_0^2 - \omega^2} - \frac{1}{2}\frac{\cos\omega_+ t - \cos\omega_0 t}{\omega_0^2 - \omega_+^2} - \frac{1}{2}\frac{\cos\omega_- t - \cos\omega_0 t}{\omega_0^2 - \omega_-^2}\right]. \end{aligned} \tag{2.78}$$

Consequently, for this last type of envelope there are potentially three types of resonances, i.e. $\omega = \omega_0, \omega_\pm$. When $\omega = \omega_0$, the displacement field (2.73) reads

$$\begin{aligned} \lim_{\omega\to\omega_0} u(t) = &\frac{q\mathscr{E}_0}{4B\omega_0^2} \\ &\left[\omega_0 t \sin\omega_0 t + \frac{\omega_0^2}{\Omega(\omega_0^2 - \Omega^2)}(\Omega \sin\omega_0 t \cos\Omega t - \omega_0 \cos\omega_0 t \sin\Omega t)\right]. \end{aligned} \tag{2.79}$$

Assuming that before the electromagnetic perturbation is switched on the oscillator is found in the fundamental state ($n = 0$), then the time-evolved wavefunction Φ reads

$$\Phi(\xi - u(t), t) = \frac{1}{\sqrt{\xi_0\sqrt{\pi}}} \exp\left[-\frac{(\xi - u(t))^2}{2\xi_0^2} - \mathrm{i}\omega_0 t\right], \tag{2.80}$$

where $\xi_0^2 = \hbar/B\omega_0$ is the square of the oscillator length.

Before calculating matrix elements of coordinate-dependent operators, $\hat{O}(\xi)$, it is worth observing that the phase function plays no role. Consequently only the real part of Φ is needed. Then, the matrix element of $\hat{O}(\xi)$ looks like

$$\langle\psi(\xi, t)|\hat{O}(\xi)|\psi(\xi, t)\rangle = \frac{1}{\xi_0\sqrt{\pi}} \int_{-\infty}^{\infty} \mathrm{d}\xi\ \hat{O}(\xi) \mathrm{e}^{-\frac{(\xi - u(t))^2}{\xi_0^2}}. \tag{2.81}$$

For example,

$$\langle\psi(\xi, t)|\hat{\xi}|\psi(\xi, t)\rangle = u(t). \tag{2.82}$$

2.3.2 The quasi-energy method

This method applies to quantum systems acted upon by an external time-periodic force such that the total Hamiltonian,

$$H = H_0(x) + H_{\text{int}}(x, t), \tag{2.83}$$

where H_0 is the unperturbed Hamiltonian possessing a complete orthonormal set of eigenfunctions $\varphi_n(x)$ and eigenvalues E_n, varies periodically in time, i.e.

$$H(x, t + T) = H(t), \qquad H_{\text{int}}(x, t + T) = H_{\text{int}}(x, t). \tag{2.84}$$

The time-dependent part of the Hamiltonian is assumed to be $H_{\text{int}}(t) \ll H_0$ and can be thus treated as an interaction which induce transitions between the eigenstates E_n of H_0. Inspired by the work of Floquet on linear differential equations with periodic coefficients [13], Zeldovich[1] [14] proposed to seek solutions of the corresponding time-dependent Schrödinger equation (TDSE) in the form

$$\psi_\alpha(t + T) = \mathrm{e}^{-\mathrm{i}\alpha}\psi_\alpha(t). \tag{2.85}$$

This *Ansatz* reflects the fact that after a period $T = 2\pi/\omega$ the wavefunction is multiplied by a phase factor. In full analogy to the concept of quasi-momentum of a particle in a spatially periodic potential of a crystal lattice (see for example [15, 16]), Zeldovich introduced the *quasi-energy*

$$\varepsilon_\alpha = \frac{\hbar\alpha}{T}. \tag{2.86}$$

With the choice (2.85), the wavefunction of quasi-energy ε_α is factorized by a periodic part,

$$\psi_\alpha(x, t) = \mathrm{e}^{-\mathrm{i}\varepsilon_\alpha t/\hbar}\phi_\alpha(x, t), \tag{2.87}$$

such that the square-integrable function ϕ_α is periodic in time (*Floquet mode*)

$$\phi_\alpha(t + T) = \phi_\alpha(t). \tag{2.88}$$

A state of the system represented by such a solution is called a *steady state* and the real-valued quantity ε_α is dubbed the *quasi-energy* of the state. The definition of the quasi-energy and the corresponding wavefunction reflects the formal analogy with the quasi-momentum characterizing Bloch functions of a crystal [16]. The TDSE satisfied by ψ reads

[1] Yakov Borisovich Zeldovich (1914–87), a Soviet physicist without higher education, addressed a wide variety of problems in many branches of theoretical physics and chemistry: nuclear physics, elementary particles, physical chemistry, theory of shock waves and combustion, physics of explosion, hydrodynamics and heat exchange, nonlinear dynamics, relativistic astrophysics, quantum mechanical systems perturbed by external fields, and mathematical physics. Between 1945 and 1963 he participated in the Soviet program for research and development of atomic and then hydrogen mass destruction weapons.

$$i\hbar\frac{\partial\psi(x,\, t)}{\partial t} = H(x,\, t)\psi(x,\, t). \tag{2.89}$$

Substituting equation (2.87) in equation (2.89) one finds the Schrödinger equation

$$\mathscr{H}\phi_\alpha(x,\, t) = \varepsilon_\alpha\phi_\alpha(x,\, t), \tag{2.90}$$

where the operator

$$\mathscr{H} \equiv H(x,\, t) - i\hbar\frac{\partial}{\partial t} \tag{2.91}$$

is the Hamiltonian for *steady states* [17]. This linear and Hermitian operator is defined over the composite Hilbert space $\mathscr{R} \otimes \mathscr{T}$ of the linear space spanned by square-integrable functions $f(x)$ on the configuration space denoted by $\mathscr{R}$ and the Hilbert space $\mathscr{T}$ of all possible functions periodic in time $g(t)$, of period T, finite

$$\int_{-T/2}^{T/2} dt\;\; |g(t)|^2, \tag{2.92}$$

and endowed with the inner product

$$(g,\, h) \equiv \frac{1}{T}\int_{-T/2}^{T/2} dt\;\; g^*(t)h(t). \tag{2.93}$$

The composite Hilbert space $\mathscr{R} \otimes \mathscr{T}$ is spanned by all possible periodic functions $\phi(x,\, t)$ with period T that make the integral

$$\int_{-T/2}^{T/2}\int dx dt\;\; |\phi(x,\, t)|^2 \tag{2.94}$$

finite. The inner product defined on $\mathscr{R} \otimes \mathscr{T}$ of two functions $\phi_\alpha(x,\, t)$ and $\phi_\beta(x,\, t)$ is

$$\langle\langle\phi_\alpha(x,\, t),\, \phi_\beta(x,\, t)\rangle\rangle \equiv \frac{1}{T}\int_{-T/2}^{T/2}\int dx dt\;\; \phi_\alpha^*(x,\, t)\phi_\beta(x,\, t). \tag{2.95}$$

Since a state vector in the $\mathscr{R} \otimes \mathscr{T}$-space can be expanded in terms of the complete orthonormal set $e^{-in\omega t}f_n(x)$, then

$$\phi_\alpha(x,\, t) = \sum_{n=-\infty}^{\infty} e^{-in\omega t}f_{n\alpha}(x), \tag{2.96}$$

and the state vector (2.87) assumes the form

$$\psi_\alpha(x,\, t) = e^{-i\varepsilon_\alpha t/\hbar}\sum_{n=-\infty}^{\infty} e^{-in\omega t}f_{n\alpha}(x). \tag{2.97}$$

The Fourier component $f_{n\alpha}$ corresponds to the state obtained from the state of energy ε_α by exchanging n photons with the laser field. Expanding H_{int} in a Fourier series,

$$H_{\text{int}} = \sum_{n=-\infty}^{\infty} e^{-in\omega t} V_n(x), \tag{2.98}$$

and substituting equations (2.97) and (2.98) in the TDSE (2.89), an infinite set of coupled time-independent equations is obtained,

$$(H_0 - n\hbar\omega)f_{n\alpha} + \sum_{m=-\infty}^{\infty} V_{n-m}f_{m\alpha} = \varepsilon_\alpha f_{n\alpha}. \tag{2.99}$$

In the case of a monochromatic laser field and assuming the validity of the dipole approximation (see chapter 1)

$$H_{\text{int}} = qx\mathscr{E}_0 \cos \omega t, \tag{2.100}$$

the set of coupled equations is transformed into

$$(H_0 - n\hbar\omega)f_{n\alpha} + \frac{1}{2}qx\mathscr{E}_0\left(f_{n-1\alpha} + f_{n+1\alpha}\right) = \varepsilon_\alpha f_{n\alpha}. \tag{2.101}$$

In the field-free limit ($\mathscr{E}_0 \to 0$) the above set is decoupled into the equations

$$(H_0 - n\hbar\omega)f_{n\alpha} = \varepsilon_\alpha f_{n\alpha}, \tag{2.102}$$

and the state $f_{0\alpha}$ can be taken as an eigenstate of the unperturbed Hamiltonian, whereas the Fourier components with $n \neq 0$ are set to 0.

Next consider the case of a quantum system with N bound states interacting with a laser field. The Fourier series coefficients $f_{n\alpha}$ are expanded in terms of an orthonormal set of eigenfunctions $\psi_m^{(0)}$ of the field-free Hamiltonian H_0,

$$f_{n\alpha}(x) = \sum_{n'=0}^{N} C_{nn'}^{(\alpha)}\psi_{n'}^{(0)}(x), \tag{2.103}$$

such that

$$H_0\psi_m^{(0)} = E_m\psi_m^{(0)}. \tag{2.104}$$

Inserting equation (2.103) in equation (2.101) and taking the overlap of the obtained equations with the eigenstate $\psi_m^{(0)}$, an infinite algebraic equations for the coefficients $C_{nm}^{(\alpha)}$ is obtained

$$(E_{n'} - \varepsilon_\alpha - n\hbar\omega)C_{nn'}^{(\alpha)} + \frac{1}{2}q\mathscr{E}_0 \sum_{m=-\infty}^{+\infty} (x)_{n'm}\left(C_{n-1,\,m}^{(\alpha)} + C_{n+1,\,m}^{(\alpha)}\right) = 0, \tag{2.105}$$

where

$$(x)_{n'm} = \int_{-\infty}^{+\infty} dx\psi_{n'}^{(0)*}x\psi_m^{(0)}. \tag{2.106}$$

2.3.3 Kramers–Henneberger method

As pointed out in the last chapter the interaction between a quantum system, such as an atom or a nucleus, can be approximately resolved in a non-inertial frame where a charged particle (electron, proton, nuclear cluster) is at rest. The Hamiltonian in the oscillating frame (Kramers frame) was written in (1.124) from where it is transparent that the dependence on the radiation field is contained in the vector $\vec{\alpha}(t)$ that shifts the origin of V [12],

$$\mathbf{r} \longrightarrow \mathbf{r} + \vec{\alpha}(t). \tag{2.107}$$

The TDSE in the Kramers frame is

$$i\hbar \frac{\partial \Psi^{\mathrm{K-H}}(\mathbf{r}, t)}{\partial t} = \left[-\frac{\hbar^2}{2\mu} \nabla_r^2 + V(\mathbf{r} + \vec{\alpha}(t)) \right] \Psi^{\mathrm{K-H}}(\mathbf{r}, t). \tag{2.108}$$

A convenient route to find the corresponding solution of the TDSE, in the case that the incident laser field comes in the form of a monochromatic plane wave (1.28), is represented by the Fourier decomposition of the K–H transformed static central potential,

$$V(\mathbf{r} + \vec{\alpha}) = \frac{1}{(2\pi)^{3/2}} \int d\mathbf{k} \ \mathcal{V}(\mathbf{k}) e^{-i\mathbf{k}\cdot(\mathbf{r}+\vec{\alpha})}. \tag{2.109}$$

Substituting in the above expression (1.96) and the Jacobi–Auger formula (A.31), which in this particular case reads

$$e^{-i\mathbf{k}\cdot\hat{\varepsilon}\alpha_0 \sin \omega t} = \sum_{n=-\infty}^{+\infty} e^{-in\omega t} J_n(\mathbf{k} \cdot \hat{\varepsilon}\alpha_0), \tag{2.110}$$

the K–H transformed central potential can be expressed as a Fourier series

$$V(\mathbf{r} + \vec{\alpha}) = \sum_{n=-\infty}^{+\infty} e^{-in\omega t} \mathcal{V}_n(\alpha_0; \mathbf{r}), \tag{2.111}$$

where the Fourier components of the potential are given by

$$\mathcal{V}_n(\alpha_0; \mathbf{R}) = \frac{1}{(2\pi)^{3/2}} \int d\mathbf{k} \ \mathcal{V}(\mathbf{k}) J_n(\mathbf{k} \cdot \hat{\varepsilon}\alpha_0) e^{-i\mathbf{k}\cdot\mathbf{r}}. \tag{2.112}$$

The perturbative method devised by Henneberger [18] consists in building an effective binding potential which is given by the $n = 0$ term in the infinite sum (2.111). Therefore, within the Henneberger formulation the eigenvalue problem of the Hamiltonian

$$H_0 = -\frac{\hbar^2}{2m} \nabla^2 + V_0(\mathbf{r}) \tag{2.113}$$

is first solved, and the next transition probabilities (ionization, tunneling) are calculated for a time-dependent perturbation specified by

$$H_{\text{int}} = \frac{1}{(2\pi)^{3/2}} \sum_{\substack{n=-\infty \\ n\neq 0}}^{+\infty} \mathrm{e}^{-in\omega t} \int \mathrm{d}\mathbf{k}\ \mathcal{V}(\mathbf{k}) J_n(\mathbf{k}\cdot\hat{\varepsilon}\alpha_0)\mathrm{e}^{-\mathrm{i}\mathbf{k}\cdot\mathbf{r}}. \tag{2.114}$$

Since for increasing intensity of the radiation, $\mathscr{E}_0 \longrightarrow \infty$, the argument of the Bessel function in the formula above, $\mathbf{k}\cdot\alpha_0 \longrightarrow \infty$, the asymptotic expansion of the Bessel function for a large argument,

$$J_n(\mathbf{k}\cdot\vec{\alpha}_0) \approx \sqrt{\frac{2}{\pi\mathbf{k}\cdot\vec{\alpha}_0}} \cos\left(\mathbf{k}\cdot\vec{\alpha}_0 - \frac{1}{2}n\pi - \frac{1}{4}\pi\right), \tag{2.115}$$

ensures that each term of the sum in (2.114) is bounded. Henneberger concluded that the merit of this method lies in the rapid convergence of the perturbation series at high intensities [18].

For small values of α_0, i.e. high frequencies/low intensities, the strategy to solve the problem in the K–H representation is to expand the oscillating potential (2.110) in powers of $\vec{\alpha}$,

$$V(\mathbf{r}+\vec{\alpha}) \simeq V(\mathbf{r}) + \vec{\alpha}\cdot\nabla V(\mathbf{r}) + \frac{1}{2}(\vec{\alpha}\cdot\nabla)^2 V(\mathbf{r}) + \cdots. \tag{2.116}$$

The most simple approximation to (2.108) consists in neglecting all Fourier components in (2.111) with $n \neq 0$. Thence, the Schrödinger equation in the oscillating frame coordinates is stationary,

$$\mathrm{i}\hbar\frac{\partial\Psi^{\text{K-H}}(\mathbf{r},t)}{\partial t} = \left[-\frac{\hbar^2}{2\mu}\nabla_r^2 + \mathcal{V}_0(\mathbf{r})\right]\Psi^{\text{K-H}}(\mathbf{r},t). \tag{2.117}$$

The static Kramers–Henneberger potential results from (2.112) for $n = 0$:

$$\mathcal{V}_0(\alpha_0;\mathbf{r}) = \frac{1}{(2\pi)^{3/2}} \int \mathrm{d}\mathbf{k}\ \mathcal{V}(\mathbf{k}) J_0(\mathbf{k}\cdot\hat{\varepsilon}\alpha_0)\mathrm{e}^{-\mathrm{i}\mathbf{k}\cdot\mathbf{r}}. \tag{2.118}$$

Due to the stationary form of (2.117) the eigenstates and eigenvalues in the Kramers–Henneberger representation can be readily obtained from

$$\left[-\frac{\hbar^2}{2\mu}\nabla_r^2 + \mathcal{V}_0(\mathbf{r})\right]\Psi_m^{\text{K-H}}(\mathbf{r},t) = E_m^{\text{K-H}}\Psi_m^{\text{K-H}}(\mathbf{r},t), \tag{2.119}$$

where m labels the energy eigenvalues $E_m^{\text{K-H}}$. This zeroth-order approximation is justified for large values of the electric field strength.

The contribution of terms with $n \neq 0$ (multiphoton effects) can be treated by means of the Floquet method, where the solution (2.108) is taken in the form

$$\Psi^{\text{K-H}}(\mathbf{r},t) = \mathrm{e}^{-\mathrm{i}Et/\hbar}\Phi(\mathbf{r},t), \tag{2.120}$$

and E is the quasi-energy and since, as discussed before, $\Phi(\mathbf{r}, t)$ is periodic, i.e. $\Phi(\mathbf{r}, t + T) = \Phi(\mathbf{r}, t)$, it can be Fourier expanded like the oscillating potential (2.111),

$$\Phi(\mathbf{r}, t) = \sum_{n=-\infty}^{+\infty} \mathrm{e}^{-in\omega t}\Phi_n(\mathbf{r}). \tag{2.121}$$

Substituting equations (2.111), (2.120), and (2.121) in equation (2.108), an infinite system of time-independent coupled equations is obtained,

$$(\mathbf{H}_\mathrm{F} - \mathrm{E}\mathbf{I})\Phi = 0. \tag{2.122}$$

Above $\mathbf{H}_\mathrm{F}$ is the matrix representation of the Floquet Hamiltonian with elements [12]

$$(\mathbf{H}_\mathrm{F})_{mm'} = -\left[\frac{\hbar^2}{2\mu}(\nabla_r^2)_{mm'} + m\hbar\omega\right]\delta_{mm'} + (\mathcal{V}_n(\alpha_0; \mathbf{r}))_{mm'}. \tag{2.123}$$

In this representation, the matrix elements of H_0 are

$$(H_0)_{mm'} = E\delta_{mm'}, \tag{2.124}$$

where E is the energy of the initial state.

The solution of equation (2.122) can be obtained by imposing appropriate boundary conditions. In the particular case of a nuclear reaction between two nuclei of charges Z_1 an Z_2, such as spontaneous emission of proton, α, or heavy clusters from the nuclear parent's ground state, or a collision such as elastic scattering, Coulomb excitation, etc, the Fourier components of the series (2.121) must be regular at the origin and can be represented asymptotically as superpositions of diverging waves,

$$\Phi_n(\mathbf{r}) \underset{r\to\infty}{\longrightarrow} \frac{1}{r}\sum_{n'} f_{nn'} \exp\left[\mathrm{i}(k_n r - \eta_n \ln 2k_n r)\right], \tag{2.125}$$

where $\hbar k_n = \sqrt{2mE_n}$ is the complex radial momentum associated with the quasi-energy of the nth channel, $E_n = E + n\hbar\omega$ and $\eta_n = Z_1 Z_2 e^2\mu/\hbar k_n$.

2.4 High-intensity regimes

2.4.1 The strong-field approximation

Strong-field methods provide a non-perturbative framework to calculate transition rates in quantum systems interacting with intense laser pulses [19]. The strong-field environment consists of the laser field that acts during a short period of time and induce transitions in a quantum system. The strategy employed by these methods consists in operating on the Schrödinger equation with a unitary transformation which approximately removes the electromagnetic field from the problem. These methods were originally designed to treat the ionization of atoms in laser fields that are strong enough to make a perturbative approach no longer adequate.

The S-matrix formalism

In non-relativistic quantum mechanics one usually considers scattering as a process where a free particle scatters under the influence of some interaction and emerges in the exit channel as a free particle with changed properties [20]. Apart from the so-called free–free processes, the *S*-matrix formalism is employed also in (quasi)bound–free, free–bound, or bound–bound processes. Examples from atomic physics are ionization, recombination, or excitation and relaxation transitions. In nuclear physics one can envisage for example: particle (proton, α, cluster) emission, radiative capture, or excitation of single-particle states or collective modes.

As a simple case study consider a charged particle of mass m, initially bound to a center of force by a finite-range potential $V(\mathbf{r})$ (electron bound to an atom, valence proton bound or quasi-bound in a nucleus), that under the action of an intense electromagnetic field is emitted in the continuum. The initial Φ_i and final (asymptotic) Φ_f states of the system, in which the electromagnetic field is turned off, form a complete set of solutions of the Schrödinger equation

$$i\hbar\frac{\partial}{\partial t}\Phi(\mathbf{r},\, t) = H_0\Phi(\mathbf{r},\, t), \tag{2.126}$$

where H_0 is the unperturbed Hamiltonian

$$H_0 = -\frac{\hbar^2}{2m}\Delta + V(\mathbf{r}), \tag{2.127}$$

and the solutions of equation (2.126) form a complete set $\{\Phi_n\}$ from which one element, Φ_i, is selected to be the initial state. In applications on atomic detachment or nuclear particle emission, the initial state wavefunction is taken to be a stationary bound state

$$\Phi_i(\mathbf{r},\, t) = \phi_i(\mathbf{r})\mathrm{e}^{-\mathrm{i}E_i t/\hbar}. \tag{2.128}$$

When the external field is turned on, the corresponding TDSE reads

$$\mathrm{i}\hbar\frac{\partial}{\partial t}\Psi = (H_0 + H_{\mathrm{int}})\Psi. \tag{2.129}$$

Above, H_{int} represents the interaction between the particle of charge q (e.g. $-e$ for an electron, e for a proton, $2e$ for an α particle) and the electromagnetic field. The solutions of (2.129) form a complete set $\{\Psi_n\}$. The condition that the laser fields acts upon the system during a finite time and in a limited space amounts to assuming the following asymptotic conditions for the vector potential $\mathbf{A}$, and *mutatis mutandis* for the interaction Hamiltonian H_{int}:

$$\mathbf{A}(t \longrightarrow \pm\,\infty) = 0, \qquad H_{\mathrm{int}}(t \longrightarrow \pm\,\infty) = 0. \tag{2.130}$$

In other words, in the remote past and distant future, when the field $\mathbf{A}$ is absent, the difference between the complete Hamiltonian and the non-interacting Hamiltonian vanish. Note that it is irrelevant to ask how the interaction is turned off at $t \longrightarrow \pm\,\infty$.

The interacting wavefunction satisfying (2.129) is initially ($t = -\infty$) chosen to coincide with the field-free state Φ_i and is called *in-state*

$$\Psi_i^{(+)}(t = -\infty) = \Phi_i. \tag{2.131}$$

Since the equations of motion of quantum mechanics are invariant under time reversal, the asymptotic condition for $\Psi_f^{(-)}$ is

$$\Psi_i^{(-)}(t = +\infty) = \Phi_f. \tag{2.132}$$

The final interacting state of the system $\Psi_f^{(-)}$—the *out-state*—will contain the complete effects of the electromagnetic field as well of the binding potential. The main task of the strong-field approach is to provide a reasonable approximation to the unknown interacting wavefunctions $\Psi_i^{(+)}$ or $\Psi_f^{(-)}$. Note that the ($\pm$) superscripts on Ψ are usually employed in scattering problems.

One must next seek the transition amplitude of the investigated process (photoatomic or photonuclear), which consists in the state Ψ turning from Φ_i (the initial state) to one of the final states Φ_f that are unbound and subjected to measurement. To that purpose the S-matrix is introduced as the overlap of the fully interacting state Ψ_i with the non-interacting state Φ_f,

$$S_{fi} = \lim_{t=+\infty} (\Phi_f, \Psi_i^{(+)}) = \lim_{t=-\infty} (\Psi_f^{(-)}, \Phi_i). \tag{2.133}$$

The asymptotic initial condition (2.131), due to the orthogonality, is equivalent to

$$\lim_{t\to-\infty} (\Phi_f(t), \Psi_i(t))) = \lim_{t\to-\infty} (\Phi_f(t), \Phi_i(t)) = \delta_{fi} \tag{2.134}$$

Combining (2.131) and (2.133), i.e. subtract from the S-matrix the no transition amplitude, and making use of the orthogonality of the $\{\Phi_n\}$ states

$$\begin{aligned} S_{fi} - \delta_{fi} &= \lim_{t=+\infty} (\Phi_f, \Psi_i^{(+)}) - \lim_{t=-\infty} (\Phi_f, \Psi_f^{(+)}) = \int_{-\infty}^{+\infty} \mathrm{d}t \ \frac{\partial}{\partial t}(\Phi_f, \Psi_i^{(+)}) \\ &= -\frac{\mathrm{i}}{\hbar} \int_{-\infty}^{+\infty} \mathrm{d}t \ \left[\left(\mathrm{i}\hbar\frac{\partial}{\partial t}\Phi_f, \Psi_i^{(+)}\right) + \left(\Phi_f, \mathrm{i}\hbar\frac{\partial}{\partial t}\Psi_i^{(+)}\right)\right]. \end{aligned} \tag{2.135}$$

The transition amplitude is obtained by making use of the equations of motion (2.126) and (2.129) in the last line of the above identity,

$$M_{fi} \equiv (S - 1)_{fi} = -\frac{\mathrm{i}}{\hbar} \int_{-\infty}^{+\infty} \mathrm{d}t \ (\Phi_f, H_{\mathrm{int}}\Psi_i^{(+)}). \tag{2.136}$$

Because of the time reversal, the same probability amplitude describes the transition from the initial state to a well-defined final state $\Phi_f = \Psi_f^{(-)}(t = +\infty)$,

$$M_{fi} = -\frac{\mathrm{i}}{\hbar} \int_{-\infty}^{+\infty} \mathrm{d}t \ (\Psi_f^{(-)}, H_{\mathrm{int}}\Phi_i). \tag{2.137}$$

This alternative is particularly useful for strong-field problems as will be discussed below.

The transition amplitude probability per unit time is given by the square of the transition amplitude divided by the time:

$$w_{fi} = \lim_{T \longrightarrow \infty} \frac{1}{T} \left| (S - 1)_{fi} \right|^2 . \tag{2.138}$$

Finally, the transition probability is obtained by integrating w_{fi} over the phase space spanned by all possible final states of the particle,

$$W = \int \frac{V \mathrm{d}\mathbf{p}}{(2\pi\hbar)^3} w_{fi}, \tag{2.139}$$

where $(2\pi\hbar)^3$ is the size of the elementary cell in phase space and V is a normalization volume.

The Reiss model of strong-field ionization

The variant of *strong-field approximation* (SFA), proposed by Reiss[2], generalizes the Keldysh method [21] to arbitrary values of the frequency and intensity of the electromagnetic field [22]. By hypothesis, the intensity of the laser field is so strong that the effect of the potential, binding the particle of mass m in the initial state E_i, can be neglected in the peripheral regions.

The S-matrix required in equation (2.135) is calculated for the interaction Hamiltonian

$$H_\mathrm{int} = -\frac{q}{m}\mathbf{p} \cdot \mathbf{A} + \frac{q^2 \mathbf{A}^2}{2m}. \tag{2.140}$$

Note that the Keldysh approximation [21] employs the length gauge where the interaction Hamiltonian is

$$H_\mathrm{int} = -\mathbf{d} \cdot \mathbf{E}(0, t). \tag{2.141}$$

With the Ansatz (2.140), equation (2.137) becomes

$$(S - 1)_{fi} = -\frac{\mathrm{i}}{\hbar} \int_{-\infty}^{+\infty} \mathrm{d}t \;\; H_\mathrm{int}(t)(\Psi_f^{(-)}, \Phi_i). \tag{2.142}$$

Since the gauge invariance is not rigorously respected for strong fields, the Reiss SFA is not yielding the same results as Keldysh SFA. The *out-state* is taken in the form of the Volkov solution (1.104),

[2] Howard R Reiss (1930–2022) was an American physicist, emeritus at the American University Washington. He made fundamental contributions to ultrafast intense laser science. In 1962 he published a seminal paper that provided a stimulus for subsequent work on the strong-field approximation. He used the non-relativistic version of his theoretical framework to predict in 1980 the above-threshold ionization (ATI), which provided a foundation for today's attosecond science.

$$\Psi_f(\mathbf{r},\, t) = \frac{1}{\sqrt{V}} e^{\frac{i}{\hbar}\left(\mathbf{p}\cdot\mathbf{r} - \frac{p^2}{2m}t - \int_{-\infty}^{t} dt' H_{\text{int}}(t')\right)}, \tag{2.143}$$

where the normalization constant is changed to $V^{-1/2}$, and V is the box normalization volume for the state in continuum.

The initial bound state of energy E_i is taken in the form of a stationary wavefunction,

$$\Phi_i(\mathbf{r},\, t) = \phi_i(r) e^{-iE_i t/\hbar}, \tag{2.144}$$

which allows the separation of time and space dependence, and the transition probability amplitude is

$$(S-1)_{fi} = \frac{i}{\hbar\sqrt{V}} \tilde{\phi}_i(\mathbf{p}) \left(\frac{p^2}{2m} - E_i\right) \int_{-\infty}^{+\infty} dt\; e^{\frac{i}{\hbar}\left(\frac{p^2}{2m} - E_i\right)t} e^{\frac{i}{\hbar}\int_{-\infty}^{t} dt' H_{\text{int}}(t')}, \tag{2.145}$$

where $\tilde{\phi}_i(p)$ is the Fourier transform of $\phi_i(r)$,

$$\tilde{\phi}_i(\mathbf{p}) = \int d\mathbf{r}\; e^{\frac{i}{\hbar}\mathbf{p}\cdot\mathbf{r}} \phi_i(r). \tag{2.146}$$

For a long pulse of a linearly polarized monochromatic field

$$\mathbf{A} = -\frac{\mathscr{E}_0}{\omega} \hat{\varepsilon} \cos \omega t, \tag{2.147}$$

where $\hat{\varepsilon}$ is real and normalized ($\hat{\varepsilon}^2 = 1$) and $\mathbf{A}$ satisfies the asymptotic condition (2.130). Introducing the notations

$$\zeta = -\frac{q\mathscr{E}_0}{m\hbar\omega^2}(\mathbf{p}\cdot\hat{\varepsilon}); \qquad z = \frac{U_\text{p}}{\hbar\omega} = \frac{q^2\mathscr{E}_0^2}{4m\hbar\omega^3}, \tag{2.148}$$

the interaction Hamiltonian (2.140) reads

$$H_{\text{int}} = \hbar\omega(-\zeta \cos \omega t + z(1 + \cos 2\omega t)), \tag{2.149}$$

whereas the new form of the Volkov wavefunction is

$$\Psi_f(\mathbf{r},\, t) = \frac{1}{\sqrt{V}} \exp\left[\frac{i}{\hbar}\left(\mathbf{p}\cdot\mathbf{r} - \frac{p^2}{2m}t - z\hbar\omega t\right) + i\zeta \sin \omega t - i\frac{z}{2} \sin 2\omega t\right]. \tag{2.150}$$

Substituting these last two expressions in (2.146):

$$(S-1)_{fi} = \frac{i}{\hbar\sqrt{V}} \tilde{\phi}_i(\mathbf{p}) \left(\frac{p^2}{2m} - E_i\right) \int_{-\infty}^{+\infty} dt\; e^{\frac{i}{\hbar}\left(\frac{p^2}{2m} - E_i + z\hbar\omega\right)t - i\zeta \sin \omega t + i\frac{z}{2} \sin 2\omega t}. \tag{2.151}$$

Using the generating function formula for GBF (see equation (A.45) from the appendix),

$$e^{i\left(-\zeta \sin \omega t+\frac{z}{2} \sin 2\omega t\right)} = \sum_{n=-\infty}^{\infty} e^{-in\omega t} J_n\left(\zeta, -\frac{z}{2}\right), \tag{2.152}$$

and of the identity [23]

$$\lim_{T\to\infty} \int_{-T/2}^{+T/2} dt\ e^{\frac{i}{\hbar}(E-E_i)t} = 2\pi\hbar\delta(E - E_i), \tag{2.153}$$

leads to a new form of (2.151),

$$(S - 1)_{fi} = \frac{2\pi i}{\sqrt{V}}\tilde{\phi}_i(\mathbf{p})\hbar\omega \sum_n (n - z) J_n\left(\zeta, -\frac{z}{2}\right)\delta\left(\frac{p^2}{2m} - E_i - (n - z)\hbar\omega\right). \tag{2.154}$$

Substituting the above result in equation (2.138), the transition probability per unit time is found to be

$$w_{fi} = \frac{2\pi\hbar}{V}|\tilde{\phi}_i(\mathbf{p})|^2\omega^2 \sum_n (n - z)^2 \left| J_n\left(\zeta, -\frac{z}{2}\right)\right|^2 \delta\left(\frac{p^2}{2m} - E_i - (n - z)\hbar\omega\right). \tag{2.155}$$

To estimate the transition rate per unit solid angle, i.e. the derivative of (2.139) with respect to the solid angle Ω, it is necessary to make use of two important properties of the Dirac δ-function, i.e.

$$\delta(a(\mathbf{p}^2 - \mathbf{p}'^2)) = \frac{1}{|a|}\delta(\mathbf{p}^2 - \mathbf{p}'^2), \qquad \delta(f(p)) = \sum_i \frac{1}{|f'(p)|}\delta(p - p_i), \tag{2.156}$$

where p_i are the simple zeros of $f(p)$. After some simple manipulations the final result is

$$\frac{dW}{d\Omega} = \frac{(2m^3\omega^5)^{1/2}}{(2\pi)^2\hbar^{3/2}} \sum_{n=n_0}^{\infty} (n - z)^2(n - z - \varepsilon_B)^{1/2}|\tilde{\phi}_i(\mathbf{p})|^2 \left| J_n\left(z^{1/2}\gamma, -\frac{z}{2}\right)\right|^2, \tag{2.157}$$

where $\varepsilon_B = -E_i/\hbar\omega$ and $\gamma = 2(n - z - \varepsilon_B)^{1/2} \sin\theta$, θ being angle the between the direction of the emitted particle and the direction of laser propagation. The sum in the formula above is running from the integer n_0, constrained by the delta function to be the smallest integer part of $z + \varepsilon_B$, to infinity. In contrast to the perturbative approach, where only the lowest order terms in n are relevant, in the SFA approach the contributions to the transition amplitude of terms with $n > n_0$ can be important for atomic processes such as the above-threshold ionization (ATI). From the definition (2.148) of z it is transparent that the minimum number of exchanged photons required by the argument of the delta function increases with the intensity, i.e. $n_0 \sim \mathcal{E}_0^2$. The impulse p contained in equation (2.157) is obviously determined by the energy conservation incorporated in the argument of the delta function

$$p = (2m\hbar\omega)^{1/2}(n - z - \varepsilon_B)^{1/2}. \tag{2.158}$$

The formula for the transition rate is a function depending strongly on the laser intensity. It also substantiates an important feature of the SFA, i.e., it handles cases with a small as well with a large number of exchanged photons. The summation over n in (2.157) is a sum over the number of absorbed photons truncated by the energy conservation.

The low-intensity limit consists in retaining the first-order terms in $\mathscr{E}_0$ and allow a single-photon exchange ($n_0 = 1$), a condition that imply $\hbar\omega > |E_i|$. Using the limit (A.49) from the appendix,

$$\lim_{z\to 0} J_n(z^{1/2}\mu,\, z\nu) = \frac{1}{2}\zeta, \tag{2.159}$$

the matrix element (2.151) acquires the form

$$(S-1)_{fi} = \frac{2\pi i}{\sqrt{V}}\tilde{\phi}_i(\mathbf{p})\frac{q\mathscr{E}_0}{2\omega}\delta\left(\frac{p^2}{2m} - E_i - \hbar\omega\right), \tag{2.160}$$

and thence (2.157) is rewritten in the form

$$\frac{\mathrm{d}W}{\mathrm{d}\Omega} = \frac{(2m^3)^{1/2}\omega}{2\pi^2\hbar^3}(\hbar\omega + E_i)^{3/2}|\tilde{\phi}_i(\mathbf{p})|^2 z\cos^2\theta. \tag{2.161}$$

Upon integration over the solid angle, the total probability of particle emission under the influence of a low-intensity field is

$$W = \frac{2}{3\pi}\frac{(2m^3)^{1/2}\omega}{\hbar^3}(\hbar\omega + E_i)^{3/2}|\tilde{\phi}_i(\mathbf{p})|^2 z. \tag{2.162}$$

The case of circular polarization of the electromagnetic field is analytically more easy to handle than the linear polarization case [19]. The circular polarization is specified by the unit vectors $\hat{\varepsilon}_1$, $\hat{\varepsilon}_2$, that span a plane orthogonal to the beam direction, and the electric field component assumes the form [12]

$$\mathbf{E}(t) = \frac{1}{\sqrt{2}}\mathscr{E}_0(\hat{\varepsilon}_1\sin\omega t + \hat{\varepsilon}_2\cos\omega t). \tag{2.163}$$

Keeping the same notation for z as in the case of linear polarization, the notation for ζ is instead slightly modified for convenience:

$$\zeta = \frac{q\mathscr{E}_0}{m\hbar\omega^2}p; \qquad z = \frac{q^2\mathscr{E}_0^2}{4m\hbar\omega^3}. \tag{2.164}$$

The interaction Hamiltonian (2.140) in these new variables is

$$H_{\text{int}} = \hbar\omega\left(\frac{\zeta}{\sqrt{2}}\sin(\omega - \varphi)t + z\right). \tag{2.165}$$

where φ is the phase angle of $\hat{\varepsilon}_1 + i\hat{\varepsilon}_2$ with respect to $\mathbf{p}$. Proceeding in a similar fashion to the linear polarization case,

$$(S-1)_{fi} = \frac{\mathrm{i}}{\hbar\sqrt{V}}\tilde{\phi}_i(\mathbf{p})\left(\frac{p^2}{2m} - E_i\right)\int_{-\infty}^{+\infty}\mathrm{d}t\; \mathrm{e}^{\frac{\mathrm{i}}{\hbar}\left(\frac{p^2}{2m}-E_i+z\hbar\omega\right)t - \frac{\mathrm{i}}{\sqrt{2}}\zeta\cos(\omega t-\phi)}. \tag{2.166}$$

It can be readily observed that in this case the simplification is due to the absence of the term $\sim z\sin 2\omega t$. Therefore one needs the generating function formula for ordinary Bessel functions (see equation (A.31) from the appendix) adapted to this case:

$$e^{-\frac{1}{\sqrt{2}}\mathrm{i}\zeta\cos(\omega t+\phi)} = \sum_{n=-\infty}^{\infty}(-\mathrm{i})^n e^{-\mathrm{i}n(\omega t-\phi)}J_n\left(\frac{\zeta}{\sqrt{2}}\right). \tag{2.167}$$

Repeating the same steps as before, the transition rate per unit solid angle for circular polarization is

$$\frac{\mathrm{d}W}{\mathrm{d}\Omega} = \frac{(2m^3\omega^5)^{1/2}}{(2\pi)^2\hbar^{3/2}}\sum_{n=n_0}^{\infty}(n-z)^2(n-z-\varepsilon_{\mathrm{B}})^{1/2}|\tilde{\phi}_i(\mathbf{p})|^2\left|J_n\left(\frac{1}{\sqrt{2}}z^{1/2}\gamma\right)\right|^2. \tag{2.168}$$

References

[1] Silin V P 1965 Parametric resonance in a plasma *Sov. Phys. JETP* **21** 1127–34

[2] Aliev Y M and Silin V P 1965 Plasma oscillations in a high-frequency electric field *Sov. Phys. JETP* **21** 601–7

[3] Hill G W 1886 On the part of the motion of the lunar perigée which is a function of the mean motions of the Sun and Moon *Acta Math.* **8** 1–36

[4] Whittaker E T and Watson G N 1962 *A Course of Modern Analysis* 4th edn (Cambridge: Cambridge University Press)

[5] Eisenberg J M and Greiner W 1988 *Nuclear Theory, Volume II: Excitation Mechanisms of the Nucleus* 3rd edn (Amsterdam: North-Holland)

[6] Gradshteyn I S and Ryzhik I M 1965 *Table of Integrals, Series and Products* 4th edn (New York: Academic)

[7] Baz' A I, Zeldovich Ya B and Perelomov A M 1971 *Rasseyanie, Reaktsii i Raspady v Nerelyativistskoi Kvantovoi Mehanike (Scattering, Reactions and Decay in Nonrelativistic Quantum Mechanics)* 2nd edn (Moscow: Nauka)

[8] Mişicu Ş 2023 Resonant enhancement of an x-ray laser electric field inside an atom *Int. J. Mod. Phys.* B **38** 2023

[9] Husimi K 1953 Miscellanea in elementary quantum mechanics *II Prog. Theor. Phys.* **9** 381–402

[10] Kerner E H 1958 Note on the forced and damped oscillator in quantum mechanics *Can. J. Phys.* **36** 371–7

[11] Collatz L 1967 *Differential Equations* (Stuttgart: Teubner)

[12] Gavrila M 1992 Atomic structure and decay in high-frequency fields *Atoms in Intense Laser Fields* ed M Gavrila (Boston, MA: Academic) pp 435–510

[13] Floquet G 1883 Sur les équations différentielles linéaires a coefficients périodiques *Ann. Sci. Éc. Norm. Supér.* **12** 47–88

[14] Zeldovich Y B 1967 The quasienergy of a quantum-mechanical system subjected to a periodic action *Sov. Phys. JETP* **24** 1006–8

[15] Kossevich A 1999 *The Crystal Lattice: Phonons, Solitons, Dislocations* (Berlin: Wiley)

[16] Ziman J M 1972 *Principles of the Theory of Solids* 2nd edn (Cambridge: Cambridge University Press)

[17] Sambe H 1973 Steady states and quasienergies of a quantum-mechanical system in an oscillating field *Phys. Rev.* A **7** 2203–13

[18] Henneberger W C 1968 Perturbation method for atoms in intense light beams *Phys. Rev. Lett.* **21** 838–41

[19] Reiss H R 1992 Theoretical methods in quantum optics: *S*-matrix and Keldysh techniques for strong-field processes *Prog. Quantum Electron.* **16** 1–71

[20] Joachain C J 1983 *Quantum Collision Theory* 2nd edn (Amsterdam: North-Holland)

[21] Keldysh L V 1965 Ionization in the field of a strong electromagnetic wave *Sov. Phys. JETP* **20** 1307–14

[22] Reiss H R 1980 Effect of an intense electromagnetic field on a weakly bound system *Phys. Rev.* A **22** 1786–813

[23] Schwartz L 1966 *Mathematics for the Physical Sciences* (Reading, MA: Addison-Wesley)

Chapter 3

Giant dipole modes induced by a high-frequency electric field

One of the most remarkable manifestations of nuclear collective motion is represented by giant resonances [1]. Among these collective modes, the giant dipole resonance (GDR), discovered in 1947 [2] in photo-fission measurements experiments of some heavy nuclei, is the most investigated one. A simple physical picture describes the excitation of GDR by an incident photon, whose electric component, constant over the nuclear size (the wavelength λ is much larger than the nuclear radius R), induces a coherent anti-phase displacement of the proton fluid with respect to the neutron fluid [3]. Goldhaber and Teller proposed a very simple model of GDR in spherical nuclei, by the reciprocal vibrations of the rigid proton and neutron distributions [4]. Inspired by the work of Siegfried Flügge [5] on the eigenmodes of a one-component nuclear liquid drop, Steinwedel and Jensen extended the model to two fluids (proton and neutron), interpenetrating each other [6]. The hydrodynamical treatment of protons oscillating in opposite phase to the neutrons provided one of the most transparent descriptions of an excitation mode in nuclei as demonstrated by further developments [7–17].

In this chapter a framework intended to describe the coupling between a high-frequency laser field and a bound proton–neutron fluid is put forward. To that purpose two-fluid nuclear hydrodynamics, endowed with nuclear Skyrme and Coulomb forces, is applied to a spherical nucleus with a well-defined surface and constant density in the interior. The displacement fields, used to derive the photo-absorption cross section, are obtained from a system of equations describing the conservation of mass and momentum of the two fluids coupled to the Maxwell equations.

3.1 Neutron–proton mixtures in an external electromagnetic field

In this section it is explained how the traditional framework of nuclear hydrodynamics [3, 8] is extended to a proton mixture interacting via Skyrme type forces [14, 16].

 © IOP Publishing Ltd 2023

First, the constituent velocities $\vec{\chi}'_{p,n}$ are introduced using the definition of the mass-current densities entering the expression of the Skyrme energy density $\mathscr{H}_{\text{Sky}}$,

$$\mathbf{j}_q = \frac{m}{\hbar}\rho_q\vec{\chi}'_q, \quad (q = p, n). \tag{3.1}$$

The kinetic energy of the proton and neutron fluids is

$$T = \frac{1}{2}m\int d\mathbf{r}(\rho_p\vec{\chi}'^2_p + \rho_n\vec{\chi}'^2_n). \tag{3.2}$$

As mentioned earlier, spin-saturated nuclei with constant densities inside the spherical nuclear radius in the ground state are considered. Consequently surface (gradient terms) and spin–orbit contributions are neglected in the nuclear part of the internal energy density. The Skyrme parametrization of the energy density, $\mathscr{H}_{\text{Sky}}$, in terms of B-coefficients and the α exponent [18], takes the following compact form:

$$\begin{aligned}\mathscr{H}_{\text{Sky}} &= \frac{\hbar^2}{2m}(\tau_p + \tau_n) + B_1\rho^2 + B_2(\rho_p^2 + \rho_n^2)\\ &+ B_3(\rho\tau - \mathbf{j}^2) + B_4(\rho_p\tau_p - \mathbf{j}_p^2 + \rho_n\tau_n - \mathbf{j}_n^2). + \left[B_7\rho^2 + B_8(\rho_p^2 + \rho_n^2)\right]\rho^\alpha.\end{aligned} \tag{3.3}$$

The above expression is a functional of the neutron and proton ($q = n, p$) local ρ_q, kinetic energy τ_q, and the mass-current $\mathbf{j}_q$ densities. The central one-body potential $\mathscr{U}_q$ is obtained by taking the functional derivative of the energy density with respect to the qth constituent density

$$\mathscr{U}_q \equiv \frac{\delta\mathscr{H}_{\text{Sky}}}{\delta\rho_q(\mathbf{r})} = \frac{\partial\mathscr{H}_{\text{Sky}}}{\partial\rho_q} - \nabla\cdot\frac{\partial\mathscr{H}_{\text{Sky}}}{\partial\nabla\rho_q} + \Delta\frac{\partial\mathscr{H}_{\text{Sky}}}{\partial\Delta\rho_q}. \tag{3.4}$$

The Coulomb contribution to the internal energy density is accounted for by the direct part of the Coulomb energy,

$$U_{\text{C}}(\mathbf{r}) = \frac{1}{2}e\rho_p(\mathbf{r})\Phi(\mathbf{r}) = \frac{e^2}{2}\rho_p(\mathbf{r})\int d\mathbf{r}\frac{\rho_p(\mathbf{r}')}{|\mathbf{r} - \mathbf{r}'|}, \tag{3.5}$$

where $\Phi(\mathbf{r})$ is the Coulomb potential.

The internal energy can be then evaluated via the integral over the whole space of $\mathscr{H}_{\text{Sky}}$ and U_{C},

$$U = \int d\mathbf{r}\big(\mathscr{H}_{\text{Sky}}(\mathbf{r}) + U_{\text{C}}(\mathbf{r})\big). \tag{3.6}$$

Next it is assumed that the neutron–proton 'plasma' is perturbed at time $t = 0$ by an external electromagnetic field of frequency ω and electric and magnetic strengths varying in time as

$$\mathbf{E}_0(t) = \hat{\varepsilon}\mathscr{E}_0\cos\omega t,\ \mathbf{B}_0(t) = (\hat{\mathbf{k}}\times\hat{\varepsilon})\frac{\mathscr{E}_0}{c}\sin\omega t. \tag{3.7}$$

In the above formula $\hat{\varepsilon}$ is the unit vector along the electromagnetic wave polarization and $\hat{\mathbf{k}}$ is the unit vector along the propagation direction such that $\hat{\mathbf{k}} \perp \hat{\varepsilon}$. The interaction of the proton component of the fluid mixture with the external electromagnetic field of strength (**E**, **B**) reads [19]

$$W_{\rm em} = e \int d\mathbf{r}\, \rho_p \vec{\chi}_p \cdot \left(\mathbf{E} + \dot{\vec{\chi}}_p \times \mathbf{B}\right), \tag{3.8}$$

where $\vec{\chi}_p$ is the proton fluid displacement field, related to the proton fluid velocity field via

$$\dot{\vec{\chi}}_p(\mathbf{r}, t) = \frac{\partial \vec{\chi}_p(\mathbf{r}, t)}{\partial t}. \tag{3.9}$$

From the microscopic point of view the coherent motion, associated to GDR, tends to be distributed via two-body scattering over the nuclear volume. On the other hand, from macroscopic point of view, the damping mechanism of this dipole collective mode can be pictured as the friction exerted by the proton fluid on the neutron one. A popular functional dependence is given by [10]

$$W_{\rm fric} = -m\Gamma \int d\mathbf{r}\, \rho_{\rm red}(\vec{\chi}_p - \vec{\chi}_n)(\dot{\vec{\chi}}_p - \dot{\vec{\chi}}_n), \tag{3.10}$$

where Γ, analogous to the bulk viscosity coefficient [20], is a constant related to the width of the giant resonance and the effective density of the fluid mixture,

$$\rho_{\rm red} = \frac{\rho_p \rho_n}{\rho_p + \rho_n}. \tag{3.11}$$

The justification of $W_{\rm fric}$ dependence on the relative proton–neutron velocity can be found in the fifth volume of the Landau and Lifshitz series [20]. It is argued that due to the motion of the proton fluid relative to the neutron fluid, the neutron fluid acquires an additional momentum leading to the appearance of a reaction force on the proton fluid proportional to $-(\dot{\vec{\chi}}_p - \dot{\vec{\chi}}_n)$. The problem of nuclear viscosity received due interest, particularly in connection with the damping of giant resonances and the dissipation in nuclear or deep-inelastic collisions of heavy-ions (see [21] and references therein).

It was pointed out by Danos and Greiner in [22] that the giant resonance damping in heavy nuclei ($A \approx 200$), based on the thermalization processes going via *doorway states*, yields a spreading width in the range $0.42\ \text{MeV} \leqslant \hbar\Gamma \leqslant 2.25\ \text{MeV}$. The description of dissipative processes, as revealed in the excitation of electric dipole and quadrupole resonances, was extended to account also for friction forces of the proton fluid with itself and neutron fluid with itself [10]. The experimental widths, $\hbar\Gamma \approx 3$ MeV for the isoscalar quadrupole state, and $\hbar\Gamma \approx 4$ MeV for the isovector 1^- (GDR) in ^{208}Pb, are fitted to a viscosity $\eta = 10^{-23}$ MeV s fm^{-3}. On the other hand, analysis of nuclear fission data based on two-body collisions [23] gives η in the range $(0.6 - 1.2) \times 10^{-23}$ MeV s fm^{-3}.

The dynamical equations governing the continuum-mechanical system of proton and neutron fluids are next derived by appealing to the Hamilton principle applied to the four-fold action integral [24, 25]

$$\delta S = \delta \int \mathrm{d}t \mathscr{L} = 0, \tag{3.12}$$

where

$$\mathscr{L} = T - U + W_{\mathrm{em}} + W_{\mathrm{fric}}. \tag{3.13}$$

A gauge term, related to the mass balance in the fluid mixture, is added to the Lagrangian (3.13), using undetermined multipliers $\lambda_{p,\,n}$ [26],

$$C = \int \mathrm{d}\mathbf{r} \sum_q \left[\frac{\partial}{\partial t}(\lambda_q \rho_q) + \nabla \cdot (\lambda_q \rho_q \dot{\vec{\chi}}_q) - \lambda_q \frac{\partial \rho_q}{\partial t} - \lambda_q \nabla \cdot (\rho_q \dot{\vec{\chi}}_q) \right]. \tag{3.14}$$

Next, a virtual variation of the action with respect to the dynamical variables ρ_q and $\dot{\vec{\chi}}_q$ is carried out. The resulting Lagrange equations corresponding to the densities are

$$\frac{\partial \mathscr{L}}{\partial \rho_q} - \frac{\partial}{\partial x_i}\left(\frac{\partial \mathscr{L}}{\partial \rho_{q,i}} \right) - \frac{\partial}{\partial t}\left(\frac{\partial \mathscr{L}}{\partial \rho_{q,t}} \right) = 0. \tag{3.15}$$

After some mathematical manipulations the following two equations are obtained:

$$\begin{aligned} &\frac{1}{2} m \dot{\vec{\chi}}_p^{\,2} - \mathscr{U}_p - e\Phi + e\vec{\chi}_p \cdot \left(\mathbf{E} + \dot{\vec{\chi}}_p \times \mathbf{B} \right) - m\Gamma \frac{\rho_n}{\rho} (\vec{\chi}_p - \vec{\chi}_n)(\dot{\vec{\chi}}_p - \dot{\vec{\chi}}_n) \\ &+ \frac{\partial}{\partial t} \lambda_p + \dot{\vec{\chi}}_p \cdot \nabla \lambda_p = 0, \end{aligned} \tag{3.16}$$

$$\frac{1}{2} m \dot{\vec{\chi}}_n^{\,2} - \mathscr{U}_n - m\Gamma \frac{\rho_p}{\rho} (\vec{\chi}_p - \vec{\chi}_n)(\dot{\vec{\chi}}_p - \dot{\vec{\chi}}_n) + \frac{\partial}{\partial t} \lambda_n + \dot{\vec{\chi}}_n \cdot \nabla \lambda_n = 0. \tag{3.17}$$

On the other hand the Lagrange equations for the proton and neutron fluid velocities are

$$m\rho_p \dot{\vec{\chi}}_p - e\rho_p \left(\vec{\chi}_p \times \mathbf{B} \right) - m\Gamma \rho_{\mathrm{red}} (\vec{\chi}_p - \vec{\chi}_n) + \rho_p \nabla \lambda_p = 0, \tag{3.18}$$

$$m\rho_n \dot{\vec{\chi}}_n + m\Gamma \rho_{\mathrm{red}} (\vec{\chi}_p - \vec{\chi}_n) + \rho_n \nabla \lambda_n = 0. \tag{3.19}$$

Discarding the quadratic terms in the velocities we obtain upon combining these last two equations

$$m\ddot{\vec{\chi}}_p = e\left[\mathbf{E} + \dot{\vec{\chi}}_p \times \mathbf{B} - \vec{\chi}_p \times (\nabla \times \mathbf{E}) \right] - \nabla \mathscr{U}_p + m\Gamma \frac{\rho_n}{\rho} (\dot{\vec{\chi}}_p - \dot{\vec{\chi}}_n), \tag{3.20}$$

$$m\ddot{\vec{\chi}}_n = -\nabla \mathscr{U}_n - m\Gamma \frac{\rho_p}{\rho}(\dot{\vec{\chi}}_p - \dot{\vec{\chi}}_n). \tag{3.21}$$

The hydrodynamical equations are next supplemented with equations relating the electromagnetic fields to the charge and current distributions of the fluid mixture:

$$\nabla \cdot \mathbf{E} = \frac{e}{\varepsilon_0}\rho_p; \quad \nabla \cdot \mathbf{B} = 0; \quad \nabla \times \mathbf{E} = -\frac{\partial \mathbf{B}}{\partial t}; \quad \frac{1}{\mu_0}\nabla \times \mathbf{B} = e\rho_p\dot{\vec{\chi}}_p + \varepsilon_0 \frac{\partial \mathbf{E}}{\partial t}. \tag{3.22}$$

In the non-perturbed state the distribution of matter is homogeneous and no flow takes place inside the nucleus, i.e.

$$\rho = \rho_p + \rho_n = \rho_0; \quad \dot{\vec{\chi}}_p = \dot{\vec{\chi}}_n = 0. \tag{3.23}$$

At $t = 0$ the p–n mixture is perturbed, making density, mean-field, velocity, electric and magnetic fields fluctuate according to

$$\rho_q \longrightarrow \rho_q^0 + \delta\rho_q, \quad \mathscr{U}_q \longrightarrow \mathscr{U}_q^0 + \delta\mathscr{U}_q, \quad \dot{\vec{\chi}}_q \longrightarrow \mathbf{v}_q. \tag{3.24}$$

$$\mathbf{E} \longrightarrow \mathbf{E}_0 + \delta\mathbf{E}, \quad \mathbf{B} \longrightarrow \mathbf{B}_0 + \delta\mathbf{B}, \tag{3.25}$$

where ρ_q denote the equilibrium densities that are much larger than the fluctuations ($\delta\rho_q \ll \rho_q$).

Since the gradient contributions of the densities are neglected in the present approach, the mean-field potentials exhibit a linear dependence on the proton and neutron density fluctuations,

$$\delta\mathscr{U}_q = G_{qq}^{(1)}\delta\rho_q + G_{qq'}^{(1)}\delta\rho_{q'}. \tag{3.26}$$

The G-coefficients are, according to [16],

$$\begin{aligned} G_{qq}^{(1)} &= 2(B_1 + B_2) + (3\pi^2)^{2/3}\rho_p^{2/3}(B_3 + B_4) \\ &+ \left\{(\alpha + 1)(\alpha + 2)B_7\rho_0^2 + B_8\left[\alpha(\alpha - 1)(\rho_q^2 + \rho_{q'}^2) + 4\alpha\rho_0\rho_p + 2\rho_0^2\right]\right\}\rho_0^{\alpha-2} \end{aligned} \tag{3.27}$$

$$\begin{aligned} G_{qq'}^{(1)} &= 2B_1 + B_3(3\pi^2)^{2/3}\rho_{q'}^{2/3} \\ &+ (\alpha + 1)(\alpha + 2)B_7\rho_0^{\alpha} + \left[\alpha(\alpha - 1)(\rho_q^2 + \rho_{q'}^2) + 2\alpha\rho_0^2\right]B_8\rho_0^{\alpha-2}. \end{aligned} \tag{3.28}$$

The equations of motion for the proton fluid (3.20) can be further simplified by ignoring non-dipole and magnetic terms. After linearization, the hydrodynamic equations of the p–n mixture are

$$\frac{\partial \mathbf{v}_p}{\partial t} = -\frac{1}{m}\nabla\delta\mathscr{U}_p + \frac{\Gamma\rho_n^0}{\rho}(\mathbf{v}_p - \mathbf{v}_n) + \frac{e}{m}(\mathbf{E}_0 + \delta\mathbf{E}), \tag{3.29}$$

$$\frac{\partial \mathbf{v}_n}{\partial t} = -\frac{1}{m}\nabla\delta\mathcal{U}_n - \frac{\Gamma\rho_p^0}{\rho}(\mathbf{v}_p - \mathbf{v}_n). \tag{3.30}$$

The equations (3.29) and (3.30) ensure the momentum-balance in the proton and neutron fluids, whereas the mass balance is governed by

$$\frac{\partial\delta\rho_p}{\partial t} + \rho_p^0\nabla\cdot\mathbf{v}_p = 0, \tag{3.31}$$

$$\frac{\partial\delta\rho_n}{\partial t} + \rho_n^0\nabla\cdot\mathbf{v}_n = 0. \tag{3.32}$$

The Maxwell equations for the perturbed fields are obtained by substituting equations (3.24) and (3.25) in equation (3.22),

$$\begin{aligned} &\nabla\cdot\delta\mathbf{E} = \frac{e}{\varepsilon_0}\delta\rho_p, \quad \nabla\cdot\delta\mathbf{B} = 0 \\ &\nabla\times\delta\mathbf{E} = -\frac{\partial}{\partial t}\delta\mathbf{B}, \quad \nabla\times\delta\mathbf{B} = \mu_0 e\rho_p^0\vec{\chi}'_p + \frac{1}{c^2}\frac{\partial\delta\mathbf{E}}{\partial t}. \end{aligned} \tag{3.33}$$

This third set of equations completes the macroscopic description of the p–n mixture.

Taking the time-derivative of the set of equations (3.31) and (3.32) and substituting inside them the set (3.29) and (3.30), the coupled wave equations for the density fluctuations are obtained. Introducing the square of the qq' matrix element of the 2×2 speed of sound matrix

$$c_{qq'}^2 = \frac{\rho_q^0}{m}G_{qq'}, \tag{3.34}$$

and the *proton 'plasma' frequency*

$$\omega_p^2 = \frac{e^2\rho_p^0}{m\varepsilon_0}, \tag{3.35}$$

substituting the set (3.29) and (3.30) inside the set of equations (3.31) and (3.32), differentiated with respect to time, the coupled set of wave equations for the density fluctuations is obtained,

$$\delta\ddot{\rho}_p = c_{pp}^2\Delta\delta\rho_p + c_{pn}^2\Delta\delta\rho_n + \frac{\Gamma}{\rho}\left(\rho_n^0\delta\dot{\rho}_p - \rho_p^0\delta\dot{\rho}_p\right) + \omega_p^2\delta\rho_p, \tag{3.36}$$

$$\delta\ddot{\rho}_n = c_{np}^2\Delta\delta\rho_p + c_{nn}^2\Delta\delta\rho_n - \frac{\Gamma}{\rho}\left(\rho_n^0\delta\dot{\rho}_p - \rho_p^0\delta\dot{\rho}_p\right). \tag{3.37}$$

By virtue of the incompressibility condition, i.e. $\rho_p + \rho_n = \rho_0$, in the ground state as well in the perturbed state, the density fluctuations are not independent,

$$\delta\rho_p = -\delta\rho_n. \tag{3.38}$$

This constraint has the consequence that the wave equations are decoupled and equation (3.36) can be rewritten as

$$\delta\ddot{\rho}_p = (c_{pp}^2 - c_{pn}^2)\Delta\delta\rho_p + \Gamma\delta\dot{\rho}_p + \omega_p^2\delta\rho_p. \tag{3.39}$$

In the case of a low-frequency laser pulse, the external electric field can be assumed constant during the life-time of the excitation (quasi-static approximation), i.e. $|\mathbf{E}(t)| \approx \mathscr{E}_0$. Under this circumstance the solution of (3.39) does not depend on the frequency of the incoming radiation ω and a harmonic time dependence on the giant dipole oscillation Ω can be safely assumed,

$$\delta\rho_p \sim \mathrm{e}^{\mathrm{i}\Omega t}. \tag{3.40}$$

Using this Ansatz to separate the time-dependent part solution of (3.39), the space-dependent solution is obtained by solving the scalar Helmholtz equation,

$$\Delta\delta\rho_p + k^2\delta\rho_p = 0, \tag{3.41}$$

where

$$\mathrm{Re}\{k^2\} = \frac{\Omega^2}{c_{pp}^2 - c_{pn}^2}\left(1 + \frac{\omega_p^2}{\Omega^2}\right), \quad \mathrm{Im}\{k^2\} = \frac{\Omega\Gamma}{c_{pp}^2 - c_{pn}^2}. \tag{3.42}$$

Due to the strong polarization induced by the electric field, the proton sphere is shifted with respect to the neutron fluid, resulting therefore the excitation of a pure dipole mode, a fact that allows a space-dependent solution in the form

$$\delta\rho_p(\mathbf{r}) = Aj_1(kr)Y_{10}(\theta, \phi), \tag{3.43}$$

where the amplitude A is determined by the boundary condition that ensures no flow takes place outside the spherical nuclear surface of radius R,

$$\mathbf{n} \cdot \frac{\partial \mathbf{v}_q}{\partial t}\bigg|_{r=R} = 0. \tag{3.44}$$

With the help of the Euler equation for the proton component (3.29), the boundary condition is transformed to

$$\left[\rho_p^0(G_{pp} - G_{pn}) + \rho_n^0(G_{np} - G_{nn})\right]\mathbf{n} \cdot \nabla\delta\rho_p\big|_{r=R} = e\rho_p^0\mathbf{n} \cdot (\mathbf{E}_0 + \delta\mathbf{E})\big|_{r=R}. \tag{3.45}$$

It was shown earlier that upon dynamic polarization due to the external field $\mathbf{E}_0$, a low electric field $\delta\mathbf{E}$ is induced in the nucleus. The case with $\delta\mathbf{E} = 0$ was discussed in the literature [3] and it applies also in the present approximation based on the Ampère law, given by the fourth equation in (3.33). By neglecting the curl of the magnetic field, $\delta\dot{\mathbf{E}} \sim \mathbf{v}_p$, and therefore

$$\mathbf{n} \cdot \delta\mathbf{E}|_{r=R} \approx 0, \tag{3.46}$$

it follows that the amplitude of the proton density vibrations relates to the strength of the external electric field $\mathscr{E}_0$:

$$A = 2\sqrt{\frac{\pi}{3}} \frac{e\mathscr{E}_0 R}{G_{pp} - G_{nn}} \left[kRj_1'(kR)\right]^{-1}. \tag{3.47}$$

Furthermore, the eigenmodes of the proton–neutron out-of-phase vibrations are given by the roots of the transcendental equation:

$$j_1'(kR) = 0. \tag{3.48}$$

In the case of an incoming pulse of high-frequency the time dependence is expressed in terms of a Fourier series as the one used to solve Hill's equation [27],

$$\delta\rho_p = \mathrm{e}^{-\mathrm{i}\Omega t} \sum_{n=-\infty}^{+\infty} \mathscr{R}_n(\mathbf{r})\mathrm{e}^{-\mathrm{i}n\omega t}. \tag{3.49}$$

The central term ($n = 0$) corresponds to the zero-photon exchange channel. Inserting the above infinite sum in equation (3.39) and equating each term of the resulting sum to zero, the wave equation of the density fluctuation for n-photon exchange reads

$$(c_{pp}^2 - c_{pn}^2)\Delta\mathscr{R}_n(\mathbf{r}) + \left[(\Omega + n\omega)^2 + \mathrm{i}\Gamma(\Omega + n\omega) + \omega_p^2\right]\mathscr{R}_n = 0. \tag{3.50}$$

Introducing the square of the wave-vector associated to the nth channel,

$$k_n^2 = \frac{(\Omega + n\omega)^2}{c_{pp}^2 - c_{pn}^2}\left(1 + \frac{\mathrm{i}\Gamma}{\Omega + n\omega} + \frac{\omega_p^2}{(\Omega + n\omega)^2}\right), \tag{3.51}$$

the Fourier amplitudes in the sum (3.49) are simply solutions of the scalar Helmholtz equation corresponding to k_n,

$$\mathscr{R}_n(\mathbf{r}) = A_n j_1(k_n r) Y_{10}(\theta, \phi). \tag{3.52}$$

Introducing the notation

$$\zeta_n^2 = \frac{k_n^2}{R^2}(c_{pp}^2 - c_{pn}^2), \tag{3.53}$$

Equation (3.51) can be then be inverted and the eigenfrequency for n exchanged photons reads:

$$\Omega_n = n\omega + \zeta_n\sqrt{1 - \left(\frac{\omega_p}{\zeta_n}\right)^2 - \left(\frac{\Gamma}{2\zeta_n}\right)^2} - \frac{1}{2}\mathrm{i}\Gamma. \tag{3.54}$$

Subsequently it is assumed that the proton velocity field can be separated into same parity longitudinal, $\nabla \times \mathbf{V}_n^{\mathrm{L}} = 0$, and transverse, $\nabla \cdot \mathbf{V}_n^{\mathrm{T}} = 0$, components [17, 28]. Thus,

$$\mathbf{v}_p(\mathbf{r}) = \mathrm{e}^{-\mathrm{i}\Omega t}\sum_{n=-\infty}^{+\infty}(\mathbf{V}_n^{\mathrm{L}}(\mathbf{r}) + \mathbf{V}_n^{\mathrm{T}}(\mathbf{r}))\mathrm{e}^{-\mathrm{i}\omega t}, \tag{3.55}$$

where

$$\mathbf{V}_n^{\mathrm{L}}(\mathbf{r}) = -\frac{\mathrm{i}}{k_n}B_n^{\mathrm{L}}\mathbf{L}_n, \quad \mathbf{V}_n^{\mathrm{T}}(\mathbf{r}) = -\frac{\mathrm{i}}{k_n}B_n^{\mathrm{T}}\mathbf{N}_n, \tag{3.56}$$

can be expressed in terms of the fundamental solutions of the Helmholtz equation [29]

$$\mathbf{L}_n = \frac{1}{\sqrt{3}}\left(j_0(k_n r)\mathbf{Y}_{10}^0 + \sqrt{2}j_2(k_n r)\mathbf{Y}_{10}^2\right), \tag{3.57}$$

$$\mathbf{N}_n = \sqrt{\frac{1}{3}}\left(\sqrt{2}j_0(k_n r)\mathbf{Y}_{10}^0 - j_2(k_n r)\mathbf{Y}_{10}^2\right), \tag{3.58}$$

The angular dependence of the above vector fields is contained in the vector spherical harmonics, $\mathbf{Y}_{\mathrm{JM}}^{\mathrm{L}}(\theta, \phi)$ [30]. Due to the constraint set by the continuity equation, the amplitude of the longitudinal component can be expressed in terms of the amplitude of the density fluctuation in each channel, i.e.

$$B_n^{\mathrm{L}} = \frac{1}{\rho_p^0}(\Omega + n\omega)A_n. \tag{3.59}$$

To resolve the fluctuations of the electromagnetic field, Ampère's law is differentiated with respect to time and one also takes the curl of the Maxwell–Faraday law, given by the third equation in (3.33), thus eliminating $\nabla \times \partial\delta\mathbf{B}/\partial t$. Eventually the non-homogeneous wave equation for the electric field fluctuations is secured:

$$\Delta\delta\mathbf{E} - \frac{1}{c^2}\frac{\partial^2\delta\mathbf{E}}{\partial t^2} = \mu_0 e\rho_p\frac{\partial\mathbf{v}_p}{\partial t} + \frac{e}{\varepsilon_0}\nabla\delta\rho_p. \tag{3.60}$$

In what follows the longitudinal/transverse decomposition for the solution of (3.60) is assumed,

$$\delta\mathbf{E}(\mathbf{r}) = \mathrm{e}^{-\mathrm{i}\Omega t}\sum_{n=-\infty}^{+\infty}(\mathbf{E}_n^{\mathrm{L}}(\mathbf{r}) + \mathbf{E}_n^{\mathrm{T}}(\mathbf{r}))\mathrm{e}^{-\mathrm{i}\omega t}, \tag{3.61}$$

where

$$\mathbf{E}_n^{\mathrm{L}}(\mathbf{r}) = C_n^{\mathrm{L}}\mathbf{L}_n, \quad \mathbf{E}_n^{\mathrm{T}}(\mathbf{r}) = C_n^{\mathrm{T}}\mathbf{N}_n, \tag{3.62}$$

By virtue of (3.60), the amplitudes of the induced electric field can be expressed in terms of the amplitudes of the proton density and velocity fluctuations,

$$C_n^{\mathrm{L}} = \left(k_n^2 - \frac{(\Omega + n\omega)^2}{c^2}\right)^{-1}\left[e\mu_0\frac{(\Omega + n\omega)^2}{k_n} - k_n\frac{e}{\varepsilon_0}\right]A_n, \tag{3.63}$$

$$C_n^{\mathrm{T}} = \left(k_n^2 - \frac{(\Omega + n\omega)^2}{c^2}\right)^{-1} e\mu_0\rho_p \frac{(\Omega + n\omega)^2}{k_n} B_n^{\mathrm{T}}. \tag{3.64}$$

If it is additionally assumed that the fluctuation of the proton velocity field is curl-less, i.e.

$$\nabla \times \mathbf{v}_p = 0. \tag{3.65}$$

Consequently, all the fluctuations of the velocity and electromagnetic fields are restricted to longitudinal components ($\sim\mathbf{L}_n$). Imposing the boundary condition (3.45) on each channel n, the same equation, that provided earlier the overtones of the density vibrations, is obtained:

$$j_1^{'}(k_n R) = 0. \tag{3.66}$$

For a selection of four Skyrme parametrizations, table 3.1 contains the characteristics related to the fundamental mode of the GDR, when $\hbar\Gamma = 2.5$ MeV. The experimental value of the corresponding first overtone is $\hbar\Omega_{\mathrm{exp}} \approx 80A^{1/3}$ MeV, which makes the older parametrization SIII predict poorly this value, while the SkX types provides a much better choice. The boundary condition selects non-vanishing components only in the $n = \pm 1$ channels of the density fluctuation (3.52),

$$A_{\pm 1} = \frac{1}{2}\sqrt{\frac{4\pi}{3}} \frac{e\rho_p R \mathscr{E}_0}{m(c_{pp}^2 - c_{pn}^2)} \left[j_1(k_{\pm 1}R) - k_{\pm 1}R j_1^{'}(k_{\pm 1}R)\right]^{-1}. \tag{3.67}$$

The electric field component of the laser pulse, oriented along the z-axis, produces a dynamic polarization of the p–n fluid mixture and therefore induces a non-vanishing dynamic dipole moment only along this direction,

$$D_z = \int \mathrm{d}\mathbf{r}\ z\delta\rho_p(\mathbf{r}) = 2\sqrt{\frac{\pi}{3}} eR^4 \sum_{n=\pm 1} A_n \frac{j_2(k_n R)}{k_n R} \mathrm{e}^{-\mathrm{i}(\Omega + n\omega)t}. \tag{3.68}$$

Next, to calculate the classical absorption cross section, the time-derivative of (3.68) is required. The next step consists in the averaging of the energy absorbed per unit time,

Table 3.1. Giant dipole resonance in some spherical nuclei for various Skyrme parametrizations, plasmonic energy, and the energy of the GDR fundamental overtone.

Interaction	$\hbar\omega_p$ (MeV)	$\hbar\Omega$ (MeV)	$\hbar\Omega \times A^{1/3}$ (MeV)
SIII	8.4	10.8	63.9
SkM	8.8	13.9	82.5
SkA	8.7	14.2	84.0
SkP	8.9	14.6	86.7

$$\sigma_{\text{abs}} \sim \frac{1}{T}\,\text{Re}\left\{\int_0^T \text{d}t \;\; \dot{D}_z(t)\mathscr{E}_0(t)\right\}. \tag{3.69}$$

To infer the final form of σ_{abs} it is necessary to generalize the derivation carried out in the field free case, as detailed in [3],

$$\sigma_{\text{abs}} = \frac{Ze^2}{m\varepsilon_0}\left(1 + \frac{c_{np}^2 - c_{nn}^2}{c_{pp}^2 - c_{pn}^2}\right)\sum_{n=\pm 1}\frac{\Gamma/c}{\zeta_n^2 - 2}\left\{\left[\Omega + n\omega + \frac{\omega_p^2 - \zeta_n^2}{\Omega + n\omega}\right]^2 + \Gamma^2\right\}^{-1}. \tag{3.70}$$

The mathematical structure of the classical absorption cross section indicates a a two-peaked bell-shaped function when $\omega \neq 0$. The high-frequency laser field exerted on the nucleus splits the one-peaked distribution in a manner analogous to the dipole strength in deformed nuclei [3]. It can be inferred from figure 3.1 that the splitting of the GDR centroid $\hbar\Omega_0 = 13.9$ MeV for the spherical heavy nucleus ^{208}Pb is distinguishable for frequencies from the γ-ray domain. Within a quantum approach, the contribution of exchanged photon numbers with $|n| \geqslant 2$ is required in calculating the sum $\sigma_{\text{abs}} = \sum_n \sigma_n$; terms displaying a resonance structure for the GDR fundamental energy Ω and the satellites $\Omega \pm \omega$, $\Omega \pm 2\omega$, … will be manifest with increasing ω. In other words, the effect of a laser pulse in a continuous wave form produces a multiresonance shape of the absorption cross section.

By using intense laser pulses with photon frequencies shifted in the γ-ray range, manifestation of nuclear collective modes, ranging from low-energy dipole modes (pygmy resonance [33]) to binary fission, can be made feasible. For example, the Ti: sapphire laser pulse with photon energy $\hbar\omega = 1.5$ eV, propagating against a ^{208}Pb

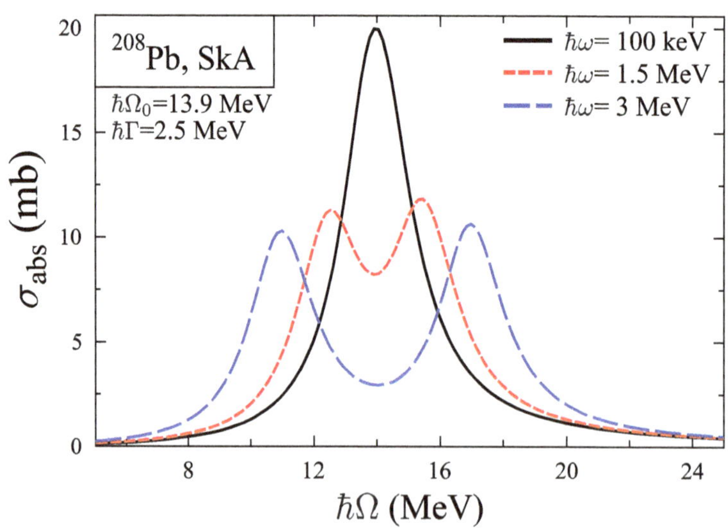

Figure 3.1. Photon absorption cross section of ^{208}Pb in a laser field for three different frequencies: $\hbar\omega = 100$ keV (full line), 1.5 MeV (short dashes), and 3 MeV (long dashes). Calculations are made for the SkM parametrization and the spreading width $\Gamma = 2.5$ MeV. Reproduced from [17]. CC BY 4.0.

beam of ultra-relativistic energy $E_{\mathrm{acc}} = 10$ TeV, is Doppler shifted with energy quanta $\hbar\omega' \approx 12.54$ keV, and experiences a large amplification of the intensity in the ion-fixed frame: $I' \approx 7 \times 10^7 I$. A beam of hard x-rays with photon energy $\hbar\omega = 9.9$ keV and intensity $I = 10^{20}$ W cm^{-2}, colliding with the same ion beam, this time of energy $E_{\mathrm{acc}} = 1$ TeV, could be a better choice. In this case $\hbar\omega' \approx 8.3$ MeV and $I' \approx 7 \times 10^{25}$ W cm^{-2}. Although the zero-photon absorption cross section displays a maximum at $\hbar\Omega = 13.5$ MeV, the energy centroid of the GDR, the one-photon σ_1, displays two maxima at 5.6 MeV and 22.3 MeV.

References

[1] Harakeh M N and van der Woude A 2001 *Giant Resonances, Fundamental High-Frequency Modes of Nuclear Excitation* (Oxford: Clarendon)

[2] Baldwin G C and Klaiber G S 1947 Photo-fission in heavy elements *Phys. Rev.* **71** 3–10

[3] Eisenberg J and Greiner W 1987 *Nuclear Theory, Volume I: Nuclear Models* (Amsterdam: North-Holland)

[4] Goldhaber M and Teller E 1948 On nuclear dipole vibrations *Phys. Rev.* **74** 1046–9

[5] Flügge S 1941 Die Eigenschwingungen eines Flüssigkeitstropfens und ihre Anwendungen auf die Kernphysik *Ann. Phys.* **431** 373–87

[6] Steinwedel H and Jensen J H D 1950 Hydrodynamik der Kerndipolschwingungen *Z. Naturforsch.* **5a** 413–20

Steinwedel H and Jensen J H D 1976 *Hydrodynamics of the Nuclear Dipole Oscillation in Photonuclear Reactions* ed E G Fuller and E Hayward (Stroudsburg, PA: Dowden, Hutchinson and Ross) (Engl. transl.)

[7] Danos M 1952 Zur Hydrodynamik der Multipolschwingungen des Atomkerns *Ann. Phys.* **10** 265–81

[8] Araújo J M 1954 The effect of nuclear compressibility on the nuclear photoeffect *Il Nuovo Cim.* **XII** 780–98

[9] Wild W 1955 Tröpfchenmodell des Atomkerns und Zweikörperkräfte *Sitzungsber. Math.-Naturw. Kl. Bayer. Akad. Wiss. München* 371–452

[10] Auerbach N and Yeverechyahu A 1975 Nuclear viscosity and widths of giant resonances *Ann. Phys.* **95** 35–52

[11] Wong C Y, Maruhn J A and Welton T A 1975 Dynamics of nuclear fluid. I. Foundations *Nucl. Phys.* A **253** 469–89

[12] Wong C Y, Welton T A and Maruhn J A 1977 Dynamics of nuclear fluid. II. Normal sound, spin sound, isospin sound, and spin–isospin sound *Phys. Rev.* C **15** 1558–70

[13] Myers W D, Swiatecky W J, Kodama T, El-Jaick L J and Hilf E R 1977 Droplet model of the giant dipole resonance *Phys. Rev.* C **15** 2032–43

[14] Krivine H, Treiner J and Bohigas O 1980 Derivation of a fluid-dynamical Lagrangian and electric giant resonances *Nucl. Phys.* A **336** 155–84

[15] Koch H, Eckart G, Schwesinger B and Holzwarth G 1982 Flow patterns of giant resonances in the first- and zero-sound approximation *Nucl. Phys.* A **373** 173–91

[16] Mişicu Ş 2016 Proton–neutron normal fluid mixtures *Rom. J. Phys.* **61** 842–56

[17] Mişicu Ş 2022 Giant dipole multi-resonances excited by high-frequency laser pulses *Particles* **5** 287–97

[18] Bonche P, Flocard H and Heenen P H 1987 Self-consistent calculation of nuclear rotations: the complete yrast line of ^{24}Mg *Nucl. Phys.* A **467** 115–35

[19] Landau L D and Lifshitz E M 1962 *The Classical Theory of Fields* (Oxford: Pergamon)
[20] Landau L D and Lifshitz E M 1987 *Fluid Mechanics* 2nd edn (Amsterdam: Elsevier)
[21] Hasse R W 1978 Approaches to nuclear fission *Rep. Prog. Phys.* **41** 1027–101
[22] Danos M and Greiner W 1965 Damping of the giant resonances in heavy nuclei *Phys. Rev.* **138** B876–91
[23] Davis K T R, Sierk A J and Nix J R 1976 Effect of viscosity on the dynamics of fission *Phys. Rev.* C **13** 2385–403
[24] Lanczos C 1970 *The Variational Principles of Mechanics* 4th edn (New York: Dover)
[25] Yourgrau W and Mandelstam S 1968 *Variational Principles in Dynamics and Quantum Theory* (New York: Dover)
[26] Eckart C 1938 The electrodynamics of material media *Phys. Rev.* **54** 920–3
[27] Whittaker E T and Watson G N 1927 *A Course of Modern Analysis* 4th edn (Cambridge: Cambridge University Press)
[28] Mişicu Ş 2006 Interplay of compressional and vortical nuclear currents in overtones of the isoscalar giant dipole resonance *Phys. Rev.* C **73** 024301
[29] Morse P M and Feshbach H 1953 *Methods of Theoretical Physics, Part II* (New York: McGraw-Hill)
[30] Rose M E 1955 *Multipole Fields* (New York: Wiley)
[31] Mocken G R and Keitel C H 2004 Bound atomic dynamics in the MeV regime *J. Phys. B: At. Mol. Opt. Phys.* **37** L275–83
[32] Apostol M and Ganciu M 2013 Coupling of (ultra-) relativistic atomic nuclei with photons *AIP Adv.* **3** 112133
[33] Bastrukov S I, Molodtsova I V, Podgainy D V, Mişicu Ş and Chang H K 2008 Elasticity of nuclear medium as a principal macrodynamical promoter of electric pygmy dipole resonance *Phys. Lett.* B **664** 258–64

Chapter 4

Laser-assisted nuclear radioactive processes

Single- and multi-electron ionization are typical examples of particle emission in a strong laser field. While this topic received considerable attention in atomic physics, the problem was not addressed until very recently in nuclear physics due to the high laser intensities required to excite nuclear states or to remove nuclear particles (nucleons, clusters, etc). Particle emission from nuclei is a typical example of a direct laser–nucleus reaction of non-resonant type [1], and the worldwide emergence of new powerful laser facilities offers a good opportunity to appraise the main features of particle emission from nuclei subjected to ultra-intense coherent electromagnetic radiation.

Various cases of nuclear particle radioactivity assisted by a strong laser field were discussed in very recent times, such as the decay of a heavy nucleus in the field of a strong laser field via α-particle [2–8] or proton emission [9, 10]. The nuclear photoeffect, i.e. the decaying nucleus absorbs a single γ-ray photon concomitantly with the action of an intense laser field, was addressed in [11, 12].

4.1 The semiclassical theory of quantum tunneling in a periodic field

The decay of a long-lived metastable state, such as the spontaneous emission of a nuclear particle, can be described by the tunneling of a wave-packet through a one-dimensional barrier. The goal of this section is to present the modification of the decay rate of a metastable state by a weak periodic field.

The quantum motion in a high-frequency field, pertaining to decay or nucleation, was considered by various authors [13–17]. In the static case, for a particle of charge q and mass m, and a trajectory specified by a single coordinate, the penetrability can be easily estimated using the WKB approximation [18]. For a time-varying barrier, the tunneling can be examined via the one-dimensional TDSE in the oscillating Kramers–Henneberger frame, derived in equation (1.124) of chapter 1 for the three-dimensional case,

doi:10.1088/978-0-7503-3199-9ch4

© IOP Publishing Ltd 2023

$$\mathrm{i}\hbar\frac{\partial\phi^{\mathrm{K-H}}(x,t)}{\partial t}=\left[-\frac{\hbar^2}{2\mu}\frac{\mathrm{d}^2}{\mathrm{d}x^2}+V(x+\alpha(t))\right]\phi^{\mathrm{K-H}}(x,t). \tag{4.1}$$

For a periodic-in-time electric field, with period $T = 2\pi/\omega$,

$$\mathbf{E}(t)=\hat{\varepsilon}\mathcal{E}_0\sin\omega t, \tag{4.2}$$

the quiver trajectory is given by (1.96) of chapter 1. In the one-dimensional case

$$\alpha(t)=\alpha_0\sin\omega t, \tag{4.3}$$

where

$$\alpha_0=-\frac{q\mathcal{E}_0}{m\omega^2}. \tag{4.4}$$

Assuming the trial solution of (4.1) to be

$$\phi^{\mathrm{K-H}}=A\exp\left[\frac{\mathrm{i}(S(x,t)-Et)}{\hbar}\right], \tag{4.5}$$

the time-dependent Hamilton–Jacobi, satisfied by the action S, is obtained by substituting equation (4.5) in equation (4.1),

$$\frac{\partial S}{\partial t}=\frac{\mathrm{i}\hbar}{2m}\frac{\partial^2 S}{\partial x^2}-\frac{1}{2m}\left(\frac{\partial S}{\partial x}\right)^2-V(x,t). \tag{4.6}$$

According to equation (A.18) in the appendix, the potential in the K–H representation can be decomposed into a static part and a time-modulated part:

$$V(x,t)=V_0(x)+\delta V(x,t). \tag{4.7}$$

One can assume a similar decomposition of S (see for example [13]),

$$S(x,t)=S_0(x)+\delta s(x,t), \tag{4.8}$$

and substituting this last expression in (4.6), the equations satisfied by S_0 and δs are

$$\frac{\partial S_0}{\partial t}-\frac{\mathrm{i}\hbar}{2m}\frac{\partial^2 S_0}{\partial x^2}+\frac{1}{2m}\left(\frac{\partial S_0}{\partial x}\right)^2+V_0(x)=0, \tag{4.9}$$

$$\frac{\partial \delta s}{\partial t}-\frac{\mathrm{i}\hbar}{2m}\frac{\partial^2 \delta s}{\partial x^2}+\frac{1}{m}\frac{\partial S_0}{\partial x}\cdot\frac{\partial \delta s}{\partial x}+\delta V(x,t)=0. \tag{4.10}$$

In the semiclassical approximation second-order derivatives are dropped-out and the solution of the equation satisfied by S_0 reads

$$S_0(x)=\mathrm{i}\hbar\int k_0(x)\mathrm{d}x, \tag{4.11}$$

where $k_0(x)=\sqrt{2m(V_0(x)-E)}$.

The modulated part of the action, δs, can be put in a form similar to the Fourier expansion of the time-dependent potential $V(x, t)$ (see equation (A.13) in the appendix),

$$\delta s(x, t) = \sum_{n \geqslant 1} (A_n e^{in\omega t} + \text{h. c.}). \tag{4.12}$$

The first-order linear differential equation satisfied by the amplitude A_n is

$$A_n' + \frac{n\omega m}{\hbar k_0(x)} A_n = \frac{im}{\hbar k_0(x)} V_n, \tag{4.13}$$

and the modulated solution reads

$$\delta s(x, t) = \frac{im}{\hbar} \sum_{n \geqslant 1} \left\{ e^{in\omega t} \int_{x_0}^{x} dx' \frac{V_n(x')}{k_0(x')} e^{-\frac{n\omega m}{\hbar} \int_{x'}^{x} \frac{dx''}{k_0(x'')}} + (\omega \to -\omega) \right\}, \tag{4.14}$$

where $(\omega \to -\omega)$ denotes the first term in the braces, where ω is taken with minus sign. In the case of barrier tunneling, x_0 is the left turning point of the barrier, where $\delta s = 0$. This last expression is an extension to the multiphoton ($n \geqslant 1$) case of the result derived in [13] in the one-photon ($n = 1$) case.

The merit of the semiclassical method is represented by the factorization of the tunneling rate in a periodic electric field. As an illustration take the simple example of a barrier with an inverted parabola shape,

$$V(x) = V_B - \frac{1}{2}\kappa(x - x_B)^2, \tag{4.15}$$

where V_B and x_B are the height and position of the barrier's top, and κ is the stiffness constant of the inverted harmonic oscillator potential. Retaining the first term in the sum (4.14) only two sidebands survive, i.e. $E \pm \hbar\omega$, and the quantum tunneling rate reads [16]

$$\Gamma \approx \Gamma_0 \exp\left[-\frac{2}{\hbar} \operatorname{Im} \delta s(x_\pm, t) \right], \tag{4.16}$$

where the turning points are given by

$$x_\pm = x_B \pm \sqrt{\frac{2V_B}{\kappa}}. \tag{4.17}$$

The tunneling rate in the absence of the periodic perturbation assumes the well-known expression

$$\Gamma_0 \sim \exp\left[-\frac{2}{\hbar} \int_{x_-}^{x_+} \sqrt{2m(V_0(x) - E)}\, dx \right], \tag{4.18}$$

and

$$\delta s(x_\pm, t) = 2mV_2 \cos\omega t \int_{x_-}^{x_+} \frac{dx'}{k_0(x')} e^{-\frac{2m\omega}{\hbar} \int_{x_-}^{x'} \frac{dx''}{k_0(x'')}}. \tag{4.19}$$

The Floquet components of the potential needed to calculate the tunneling rate are

$$V_0(x) = V(x) - \frac{1}{4}\kappa\alpha_0^2; \quad V_2(x) = \frac{1}{8}\kappa\alpha_0^2. \tag{4.20}$$

4.2 The modification of nuclear particle decay rates in x-ray laser fields

4.2.1 Laser-assisted alpha-decay in a one-dimensional approach

Soon after the discovery of natural radioactivity, attempts were made to modify the decay rates of α-radioactivity. Rutherford proposed such an experiment, where a sudden change of temperature, up to 2500 °C, and pressure, up to 100 MPa, for a short period of time were applied to an α-active probe. No effect was observed in the measured half-lives (as cited in [19]) and this negative result was interpreted in the Gamow[1] theory of quantum-mechanical tunneling. It was argued that induced distortions or even removal of the electron cloud surrounding the mother nucleus, induced by temperature, pressure, magnetic or gravitational fields, produce a negligible effect on the Coulomb barrier. Consequently, changes in the decay rate are estimated to be of the order $\delta\lambda/\lambda \approx 10^{-7}$ [20]. In more recent times it has been suggested that α decay of transuranic nuclear waste can be stimulated by embedding in it metals at low temperature [21]. This proposal was soon refuted based on the grounds that medium modifications are disturbing the alpha-daughter potential mainly at radii larger than the outer turning point [22].

In this section the effect of a linearly polarized, ultra-intense laser field on the tunneling probability of an α-particle emitted from a heavy spherical nucleus is investigated. For that purpose, a non-perturbative formalism adequate for strong fields is employed. The TDSE describing the relative alpha cluster—daughter nucleus motion under the effect of a continuous wave (CW) or a sequence of ultrashort pulses containing a few even or odd numbers of half-cycles of duration no longer than 1 as, is solved by direct numerical integration. The latter issue is related to the opportunity, available after the turn of the millennium, to access parameter ranges in high-field physics by using laser pulses comprising a small number of cycles [23]. Phase-controlled light pulses are expected to allow the control of high-intensity light–matter interactions on a sub-cycle time scale. The primary goal is to determine how the interaction of radiation with the metastable α-daughter system alters the decay rates, i.e. decide whether the decay is speed-up or hindered.

Below the α-decay of the nucleus the ^{106}Te case is discussed. This nucleus was subjected to experimental investigations in recent times along with other α-emitters slightly above the double-magic nucleus ^{100}Sn [24]. The measurement of α-decay energies of trans-tin emitters indicates an increase along the decay chains as a result of particularly strong binding of nuclei with $Z \approx N$ in this mass region. This

[1] George Gamow (1904–68) former Soviet physicist who defected to the West in 1933, is known for the famous theory of α-decay which explained the Geiger–Nuttall law. He was the author of many books of popular physics.

conclusion was confirmed by penetrability calculations in the framework of the WKB approximation [25].

Among the methods developed at the atomic level some of them could be good candidates as tools in modeling the laser–nucleus interaction [26]. If one takes for example a laser pulse with a peak of the electric field much smaller than the electric fields experienced by an α-cluster on the surface of a heavy nucleus, the theoretical treatment can be done by resorting to perturbative methods. Since in this book ultra-intense laser fields are considered, the problem is tractable in the framework of non-perturbative methods that require a direct numerical integration of the Schrödinger equation. These methods also have the advantage that solutions can be obtained for a wide range of laser intensities and frequencies, regardless of the pulse type.

The treatment of α-decay in ultra-intense laser fields is treated in what follows in a quantum-mechanical approach under the proviso of two simplifying assumptions: the electric field is assumed to be linearly polarized and its direction $\mathbf{E}(t) = \mathscr{E}(t)\mathbf{e}_x$, taken along the x-axis, is parallel to the symmetry axis of the α-daughter system. Thus, a fixed orientation in space of the decaying system is selected and all possible non-axial configurations with respect to the field are excluded. In the next section the proton decay will be treated in a two-dimensional extension of the same approach used in this section for α-decay. In the one-dimensional approximation the TDSE describing the tunneling process in the *length gauge* (1.119) reads

$$i\hbar\frac{\partial\phi(x,t)}{\partial t} = \left[-\frac{\hbar^2}{2\mu}\frac{\mathrm{d}^2}{\mathrm{d}x^2} + V(x) + H_{\mathrm{int}}(x,t))\right]\phi(x,t). \tag{4.21}$$

This equation describes the one-dimensional quantum motion of an α-particle inside a static potential $V(x)$, subjected during the tunneling process to an additional time-dependent external field. As can be inferred from equation (1.116)

$$H_{\mathrm{int}} = -eZ_{\mathrm{eff}}x\mathscr{E}_0(t), \tag{4.22}$$

where $Z_{\mathrm{eff}} = (Z_1A_2 - Z_2A_1)/(A_1 + A_2) \approx 0.98$, since in the discussed case $Z_1 = 2$, $Z_2 = 50$, and $A_1 = 4$, $A_2 = 102$. The static α-daughter $V(x)$ potential is the sum of a long range Coulomb potential [27]

$$\begin{aligned} V_{\mathrm{C}}(x) &= \frac{Z_1Z_2e^2}{8\pi\varepsilon_0 x}\left(3 - \frac{x^2}{R_{\mathrm{C}}^2}\right), \quad x < R_{\mathrm{C}}, \\ &= \frac{Z_1Z_2e^2}{4\pi\varepsilon_0 x}, \quad x \geqslant R_{\mathrm{C}}, \end{aligned} \tag{4.23}$$

where $R_{\mathrm{C}} = 1.2 \times A_2^{1/3}$ is the charge radius, and

$$V_{\mathrm{nucl}} = \frac{V_0}{1 + e^{(|x| - R_{\mathrm{N}})/a}}, \tag{4.24}$$

the nuclear potential, where V_0 is the depth of the Woods–Saxon well, $R_{\mathrm{N}} = r_0 \times A_2^{1/3}$ ($r_0 = 1.235$ fm) is the nuclear radius, and a is the potential diffuseness.

The electric field of a non-dispersive laser can be represented by a modulated linear-polarized and monochromatic plane wavefunction (single-mode field)

$$E(t) = \mathscr{E}_0 F(t)\sin(\omega t + \delta), \tag{4.25}$$

where $\mathscr{E}_0$ is the amplitude of the electric field strength, $F(t)$ is the pulse shape function (envelope), and $\omega = 2\pi/T$ is the carrier angular frequency. Since $\mathscr{E}_0$ is the peak of the pulse, $F(t)$ is taken to vary between 0 and 1. The carrier-envelope phase (CEP) δ is the phase of the carrier wave with respect to the maximum $F(t)$ and it is meaningless in what concerns the response of the decaying system when the pulses are long. It is also assumed that a maximum of the laser pulse envelope occurs at $t = 0$.

A few comments concerning equation (4.21) deserve to be made. Since the dipole interaction H_{int} is proportional to x, numerical instabilities at the computational boundaries are expected. This term can be removed by converting the Hamiltonian using the transformation [28]

$$\phi(x, t) = \exp\left(-\frac{\mathrm{i}}{\hbar} e Z_{\text{eff}} A(t) \cdot x\right)\tilde{\phi}(x, t). \tag{4.26}$$

The transformed TDSE,

$$\mathrm{i}\hbar\frac{\partial\tilde{\phi}}{\partial t} = \frac{1}{2\mu}\left[\left(\mathrm{i}\hbar\frac{\partial}{\partial x} - eZ_{\text{eff}}A(t)\right)^2\tilde{\phi}\right] + V(x)\tilde{\phi}, \tag{4.27}$$

is solved by direct numerical integration using the method of Crank–Nicolson (CN) [29]. The formal solution of equation (4.27) is

$$\tilde{\phi}(x, t + \Delta t) = \mathrm{e}^{-\frac{\mathrm{i}\Delta t}{\hbar}H(x, t)}\tilde{\phi}(x, t), \tag{4.28}$$

where $H(x, t)$ is the Hamiltonian given by

$$H = \frac{1}{2\mu}\left(\mathrm{i}\hbar\frac{\partial}{\partial x} - eZ_{\text{eff}}A(t)\right)^2 + V(x). \tag{4.29}$$

Expanding H in Taylor series up to the second-order in t and using the Padé approximation, the exponential in equation (4.28) is expressed as the ratio of two first-order polynomials (the Cayley transformation)

$$\mathrm{e}^{-\frac{\mathrm{i}\Delta t}{\hbar}H} \approx \frac{1 - \frac{\mathrm{i}\Delta t}{2\hbar}H - \frac{\mathrm{i}\Delta t^2}{4\hbar}\dot{H}}{1 + \frac{\mathrm{i}\Delta t}{2\hbar}H + \frac{\mathrm{i}\Delta t^2}{4\hbar}\dot{H}}, \quad \dot{H} = \frac{\partial H}{\partial t}. \tag{4.30}$$

In this manner the solution $\tilde{\phi}^{n+1}$ at the moment $t + \Delta t$ is obtained from the solution at the moment t, $\tilde{\phi}^n$, with an error in time proportional to Δt^3, according to the formula

$$\left(1 + \frac{i\Delta t}{2\hbar}H + \frac{i\Delta t^2}{4\hbar}\dot{H}\right)\tilde{\phi}^{n+1} = \left(1 - \frac{i\Delta t}{2\hbar}H - \frac{i\Delta t^2}{4\hbar}\dot{H}\right)\tilde{\phi}^{n}. \tag{4.31}$$

The derivatives with respect to x appearing in H and $\dot{H}$ are in practice approximated by finite differences on spatial mesh points and the solution at each time step is obtained by solving a linear system.

The application of the CN scheme presents a series of advantages: the initial wavefunction is required only at the starting value of time, it is unconditionally stable, it is unitary and conserves the norm [30], and, by approximating the derivatives with respect to x by the usual three-point formulas, at each time step a tridiagonal linear system should be solved; this can be done quickly and accurately up to the machine precision.

It should also be pointed out that when the first derivative with respect to x is absent in the Hamiltonian, the spatial accuracy can be improved either by using Numerov-like formulas [31, 32], or by approximating the second derivative with higher order finite difference formulas [33]. In the latter case, more complicated linear systems result, with the number of diagonals larger than 3, during the time propagation. In the present case, both derivatives, i.e. first and second, appear in the Hamiltonian and in order to increase the spatial accuracy, finite difference formulas in more points should be used for each derivative, leading to non-tridiagonal systems which require longer computational effort at each time step. The accuracy in time could also be increased by using the Magnus expansion and diagonal Padé approximation [34] with the cost that additional computing time is required. For example, for a scheme of order $(\Delta t)^5$, a pair of linear systems should be solved in order to advance one step in time [35].

In view of the high laser intensities considered in this problem, the emitted cluster can acquire a quiver motion $\alpha(t)$ much larger than the initial average size of the dinuclear system and therefore the forerunners of the wavefunction can travel large distances in short times. Under these circumstances the spatial grids used in the integration scheme must be correspondingly large and posses a short step in order to control as good as possible the fine changes in the wavefunction. For the sake of accuracy the time discretization requires also a large number of small steps. Depending on the practical conditions imposed by the initial wavefunction or the laser field parameters, a spatial mesh size $\Delta x = 1/8$ fm and time step $\Delta t = 1/8$ in units of 10^{-22} s proved to be a satisfactory choice for most calculations.

Since the computation of the numerical solution has to be confined to a finite domain, say $\mathscr{D} = [x_{\text{left}}, x_{\text{right}}]$, instead of the entire real line $(-\infty, \infty)$, artificial boundary conditions are introduced. The initial data are supported on this finite domain and therefore the exact solution on the whole real axis problem can be approximated by solving the TDSE only on $\mathscr{D}$, with artificial boundary conditions imposed at x_{left} and x_{right}. When the approximate solution coincides on $\mathscr{D}$ with the exact solution, the boundary conditions are known in the literature as *transparent boundary conditions* (TBC). A proper implementation of this algorithm is paramount in order to avoid reflections of the propagated wavefunction at the grid frontiers and thus causing errors in the calculation of physical quantities. The TBC algorithm is used in the form suggested in [36] and for its implementation some important relations should be recalled, as follows.

Any solution of the TDSE (4.27) satisfies the continuity equation for the probability density:

$$\frac{\partial \rho}{\partial t} + \frac{\mathrm{d}}{\mathrm{d}x} J = 0, \tag{4.32}$$

where ρ is the probability density

$$\rho(x, t) = |\tilde{\phi}(x, t)|^2, \tag{4.33}$$

and J is one-dimensional current or probability flux

$$J(x, t) = -\frac{\mathrm{i}\hbar}{2\mu}\left[\tilde{\phi}^*(x, t)\frac{\mathrm{d}}{\mathrm{d}x}\tilde{\phi}(x, t) - \tilde{\phi}(x, t)\frac{\mathrm{d}}{\mathrm{d}x}\tilde{\phi}^*(x, t)\right] + \frac{eZ_{\mathrm{eff}}}{\mu}A(t)\rho(x, t).$$

The last two forms can be checked by direct calculation.

The time-dependent tunneling probability measuring the escape likelihood of the α-particle from the nuclear + Coulomb potential well is defined by

$$P_{\mathrm{tun}}(t) = \left(\int_{x_b}^{\infty} + \int_{-\infty}^{-x_b}\right)\rho(x, t)\mathrm{d}x = 1 - \int_{-x_b}^{x_b} \rho(x, t)\mathrm{d}x, \tag{4.34}$$

where $\pm x_b$ are the positions of the static barrier. Taking the time derivative of (4.34) and using the continuity equation, a relation between the tunneling rate and the flux across the nuclear surface is obtained,

$$\dot{P}_{\mathrm{tun}}(t) = J(x_b, t) - J(-x_b, t). \tag{4.35}$$

By the same token, introducing the total norm inside the numerical grid, i.e. integrate the probability density between the left, x_{left}, and the right x_{right} boundaries:

$$P_{\mathrm{int}}(t) = \int_{x_{\mathrm{left}}}^{x_{\mathrm{right}}} \rho(x, t)\mathrm{d}x, \tag{4.36}$$

the flux across the numerical grid results in the form

$$\dot{P}_{\mathrm{int}}(t) = J(x_{\mathrm{left}}, t) - J(x_{\mathrm{right}}, t). \tag{4.37}$$

The treatment of the two boundaries is identical, and therefore the subsequent considerations are continued by focusing only on the right boundary. Following [36], an essential assumption is made, i.e. near this boundary $\tilde{\phi} = A\exp(\mathrm{i}k_x \cdot x)$, where A and k_x are complex constants. Adopting this assumption, the flux leaving the right boundary x_{right} is

$$J(x_{\mathrm{right}}, t) = \frac{1}{\mu}[\mathrm{Real}(\hbar k_x) + eZ_{\mathrm{eff}}A(t)]|\tilde{\phi}(x_{\mathrm{right}}, t)|^2. \tag{4.38}$$

An important feature of this procedure is that k_x is allowed to change as the problem progresses, thus eliminating the need for a problem-dependent adjustable parameter. The above mentioned assumption is compatible with the following equations:

$$\frac{\tilde{\phi}(x_{M+1}, t_n)}{\tilde{\phi}(x_M, t_n)} = \frac{\tilde{\phi}(x_M, t_n)}{\tilde{\phi}(x_{M-1}, t_n)} = \exp(\mathrm{i}k_x\Delta x), \tag{4.39}$$

where x_M is the boundary of the spatial grid. The second equality delivers k_x and, if substituted in the first equality of the above equation it results that

$$\tilde{\phi}(x_{M+1}, t_n) = \tilde{\phi}(x_M, t_n)\exp(\mathrm{i}k_x\Delta x). \tag{4.40}$$

Assuming the same relation is valid for the next time step, then

$$\tilde{\phi}(x_{M+1}, t_{n+1}) = \tilde{\phi}(x_M, t_{n+1})\exp(\mathrm{i}k_x\Delta x). \tag{4.41}$$

Formula (4.38) shows that an outgoing or ingoing flux develops across the boundary depending on the sign of $\mathrm{Real}(\hbar k_x) + eZ_{\mathrm{eff}}A(t)$. Due to the action of the time-dependent field, the real part of k_x is no longer constrained to be always positive, as happens for α-decay in the absence of an external perturbation [37]. For this last process the overall change in energy from the right boundary is always negative and thus the wavefunction flows out of the grid region. In the present case, back-flow of the wavefunction inside the numerical domain is expected to occur due to the reversal of the electric field polarity. In practice the outgoing flux can be computed and add it to the norm if a part of the wavefunction leaks outside the domain. Employing the TBC procedure one can obtain values of the physical quantities with a much smaller extension of the spatial grid than is necessary without the TBC procedure.

An initial wavefunction is generated at $t = 0$ in order to solve the TDSE. This task can be accomplished by appealing to the proposal put forward a long time ago by Jackson and Brown [38]. Their recipe provides this initial wavefunction, $\phi_0(x)$, as a bound state of the stationary Schrödinger equation in the modified static potential. Remember that the potential $V(x)$ has a constant value $V(\pm x_{\mathrm{mod}}) > E_\alpha$ for a distance $|x| > |x_{\mathrm{mod}}|$ beyond the top of the barrier. Accordingly, the initial wavefunction is found by considering the stationary Schrödinger equation without any perturbation ($\mathscr{E}_0 = 0$),

$$\frac{\mathrm{d}^2\phi}{\mathrm{d}x^2} + \frac{2\mu}{\hbar^2}[E_\alpha - V_{\mathrm{mod}}(r)]\phi = 0, \tag{4.42}$$

where $V_{\mathrm{mod}}(r)$ is the modified potential. The above equation is solved by discretizing the space coordinate with the same difference formula for the second derivative and thus transforming it to a matrix eigenvalue problem. The depth V_0 of the WS-potential is varied until one of the eigenvalues is matching the experimental decay energy E_α.

Many authors highlighted the common feature of many unstable quantum systems, not only of α emitters, which consists in the tendency of the decay rate to deviate strongly from the exponential law at early times of the quantum tunneling process (see for example [39, 40]). This circumstance can be understood by appealing to the following line of argumentation: it was pointed out earlier that the wave-function at $t = 0$ is prepared as a bound eigenstate of the modified potential, and it is therefore a wave-packet with a sharp energy in contrast to the wave-packet at $t > 0$, which acquires an energy width $\Delta E \neq 0$ due to the interaction responsible for the decay. After the insertion, at $t = 0$, of the initial wavefunction in the real Hamiltonian, there will be a transient period necessary for the wave-packet to adapt to its new environment. In this transient time, τ_{tr}, much shorter than the decay time, the decay is expected to be non-exponential. In [37] this behavior was proved numerically and it was concluded that if the decay energy decreases, the time necessary for the decay rate to reach its exponential behavior regime asymptotically increases. At later times ($t > \tau_{tr}$) the probability density inside the numerical grid (4.36) obeys the radioactive decay law

$$P_{\text{int}}(t) = P_{\text{int}}(0)e^{-\lambda t}. \tag{4.43}$$

Taking the time derivative of (4.43) and inserting equation (4.37), the decay rate is obtained in terms of the quantum flux balance between the region inside the numerical grid and the exterior domain.

$$\lambda(t) = -\frac{1}{P_{\text{int}}(t)}(J(x_{\text{left}}, t) - J(x_{\text{right}}, t)). \tag{4.44}$$

In all numerical experiments presented below, $\omega\tau_{tr} < 2\pi$, i.e. the laser wave period is larger than the transition time.

It is instructive to borrow from atomic physics the so-called 'ionization' probability,

$$P_{\text{ion}} = 1 - \left| \int_{x_{\text{left}}}^{x_{\text{right}}} \tilde{\phi}^*(x, t)\phi_0(x) \right|^2. \tag{4.45}$$

This quantity provides a measure of the tunneling speed [28]. In a nuclear physics context it can be interpreted as a measure of the departure of the time-dependent α particle metastable state, $\tilde{\phi}(x, t)$, from the initial bound state $\phi_0(x)$. It should be emphasized at this point that the Gamow-like solution of the corresponding stationary Schrödinger equation with purely outgoing boundary conditions, $\psi(x, t)$, has a complex eigenvalue, $E_\alpha - i\hbar\lambda/2$ and decays exponentially in time [41]:

$$L(t) = \left| \int_{x_{\text{left}}}^{x_{\text{right}}} \psi^*(x, t)\psi(x, 0) \right|^2 = \exp[-\lambda t] \tag{4.46}$$

The decay rate λ appearing in the above formula coincides with the asymptotic value of (4.44), i.e. $\lambda = \lambda(\infty)$. For ^{106}Te, the parameters of the α-nucleus optical potential are taken from the compilation [42]: $V_0 = -137.7$MeV, $a = 0.76$ fm, and $r_0 = 1.235$ fm. They are consistent with the energy $E_\alpha = 4.15$ MeV if one solves the stationary Schrödinger equation (4.42) when the potential is modified at $|x_{\text{mod}}| = 25$ fm. The spatial border is specified by $x_{\text{right,left}} = \pm 192$ fm.

As an envelope function, a sequence of square pulses is chosen,

$$F(t) = \sum_{p=0}^{N} (-)^p \theta(t - \tau_p). \tag{4.47}$$

It acts in the time intervals $[\tau_0 = 0, \tau_1]$, $[\tau_2, \tau_3]$, etc. The CW case is reproduced when $\tau_0 \longrightarrow \tau_p$, where τ_p is the duration of a single long pulse with constant amplitude. The general case deals with constant amplitude short pulses of length equal to an integer number of half-cycles and separated by intervals of similar length. If at the time τ_{2i} the field is again turned on and then at τ_{2i+1} is again turned off, the duration of the $(i + 1)$th short pulse is

$$\tau_{i+1} - \tau_i = n_{i+1}\frac{T}{2}, \tag{4.48}$$

where n_{i+1} is the number of half-cycles.

The effect of the radiation on the dinuclear system is first analyzed for a continuous sinusoidal pulse. A comparison between the tunneling probability in the free case and in the free–free case is presented in figure 4.1. The laser field is characterized by the electric field strength of $\mathscr{E}_0 = 10^{14}$ V cm^{-1} ($e\mathscr{E}_0 = 10$ eV fm^{-1}), which corresponds to the intensity $I_0 = 1.33 \times 10^{25}$ W cm^{-2}, just slightly above the ELI maximum value, and frequency $\omega \approx 4.6 \times 10^{17}$ s^{-1} ($\hbar\omega \approx 0.3$ keV) which, translated in wavelength, is $\lambda \approx 41.3$ Å. It can be inferred that even at a rather low field intensity, compared to the barrier ($e\mathscr{E}_0 \ll V_{\text{barrier}}$), the escape chances of the α-particle are increasing. On a long term the nuclear α-daughter dipole is oscillating around an average value that increases with a very small slope.

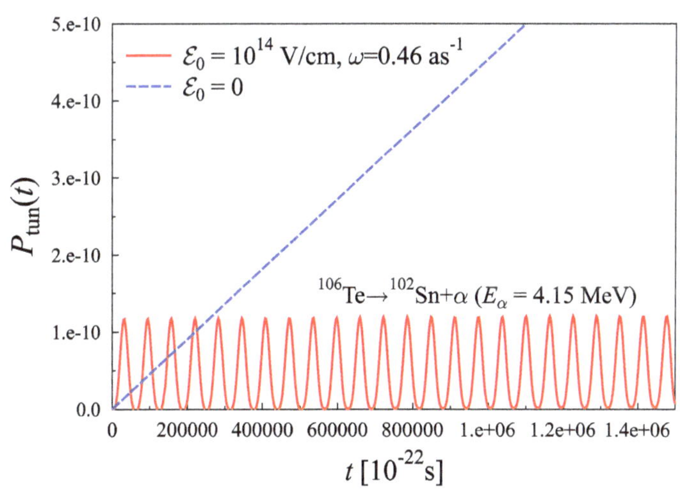

Figure 4.1. Tunneling probabilities of the α-particle from ^{106}Te for a continuous pulse of frequency $\omega = 0.46$ as^{-1} and amplitude $\mathscr{E}_0 = 10^{14}$ V cm^{-1} ($e\mathscr{E}_0 = 10$ eV fm^{-1}) and when the laser field is turned off (blue dashed curve). Reproduced from [3]. Copyright IOP Publishing Ltd.

It should be recalled that there is a phenomenon which resembles the one studied in this section, the well-known ionization of an electron moving in a static Coulomb potential and an intense laser field [43, 44]. The effect of the varying electric field is to lower by over half of a cycle the barrier seen by the electron and thus enhance the ionization probability by escape over the *downhill* barrier. In analogy to the case of atomic ionization one can speak of a stabilization of the decaying system due to the action of the radiation field.

The effect of increasing $\mathcal{E}_0$ for a fixed frequency can be visualized in figure 4.2. To that end the tunneling probabilities for the field-free case are compared to the case when a CW laser with the same frequency as in figure 4.1 acts on the decaying system with various intensities. This figure indicates the fact that the higher the intensity, the higher is the oscillation amplitude of $P_{\text{tun}}(t)$. From the classical point of view this effect is related to the oscillation of the alpha particle in the varying electric field. In this picture recollision happens when $P_{\text{tun}}(t)$ approaches its minima.

Consider now a laser field perturbing the decaying system as a sequence of abruptly turned-on short pulses consisting of an odd number of half-cycles. The tunneling probability for a laser signal formed of four pulses, each of length $3\pi/\omega$, and for the field-free case is displayed in figure 4.3. A huge field of amplitude $\mathcal{E}_0 = 10^{17}$ V cm^{-1} ($e\mathcal{E}_0 = 10$ keV fm^{-1}) and a smaller wavelength of $\lambda = 4.13$ Å are used in this case with the particular intent to give the reader a clear picture of the tunneling probability enhancement. As can be seen in the right panel of this figure, which represents the large time scale behavior, the factor gained by $P_{\text{tun}}(t)$ when applying an odd number of half-cycles is ~850 after ~150 as.

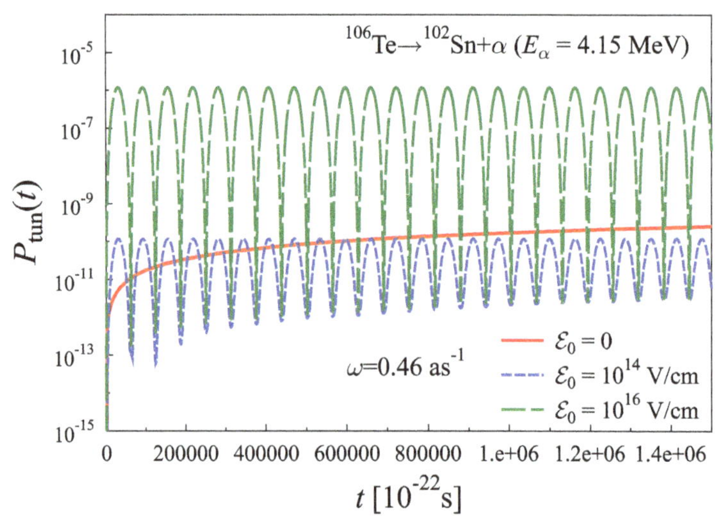

Figure 4.2. Tunneling probabilities in the logarithmic scale for two different CW field amplitudes $\mathcal{E}_0 = 10^{14}$ V cm^{-1} and 10^{16} V cm^{-1} and frequency $\omega = 0.46$ as^{-1} are compared to the tunneling probability without a laser field. Reproduced from [3]. Copyright IOP Publishing Ltd.

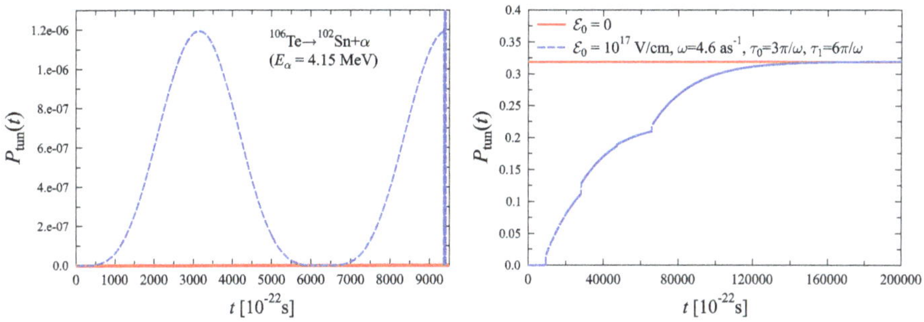

Figure 4.3. The tunneling probabilities of the α-particle from ^{106}Te for $\mathcal{E}_0 = 10$ keV fm^{-1} and $\omega = 4.6$ as^{-1} for a sequence of four pulses of duration $3\pi/\omega$ (separated by three breaks of the same duration) is compared to the field-free case. On the left panel the time axis runs up to 950 zs. Note that after a complete cycle ($2\pi/\omega$) the tunneling probabilities are similar. In the right panel the curves are represented in the time interval extended to 20 as and the field-free $P_{tun}(t)$ is multiplied by a factor of 850. Reproduced from [3]. Copyright IOP Publishing Ltd.

A suggestive representation is given in figure 4.4 of how the decay rates (4.44) are step-wise increased when the electric field is turned off. This behavior is caused by the box-like shape of the laser signal envelope. For a Gaussian or sin^2 envelope a smooth increase of $\lambda(t)$ should be expected. In this figure the pulse with an odd number of half-cycles is compared to the pulse containing an even number of half-cycles. In the second case the jump of λ is evidently smaller compared to the first case when the laser field is turned off. Since the half-life $T_{1/2}$ is inversely proportional to the asymptotic value of λ, it can then be inferred from figure 4.4 that in the most favorable case, when a very short pulse with an odd number of half-cycles is interacting with the decaying system, $T_{1/2}$ decreases by ~9 orders of magnitude. It can be therefore concluded that the decay can be speed-up in the order of femtoseconds instead of microseconds for the α-radioactive nucleus ^{106}Te.

This jump can be understood by revisiting the decay rate formula (4.44): the probability to find the particle inside the numerical domain at a given time, present in the denominator of the fraction in the rhs of this equation, displays an insignificant variation at early times, especially for a large integration domain. On the other hand the wave-packet at time $t = 0$ is confined in the nuclear domain. Thereupon, when the pulse with an odd number of half-cycles is turned off, the decay rate exhibits an abrupt variation entirely due to the imbalance in the flux. The jump is even more conspicuous when the total flux leaving or returning inside the nuclear surface is evaluated taking as reference the left and right Coulomb barrier positions. This imbalance between the flux at the left border and the one at the right border is caused by the disruption of the pulse in the middle of the oscillation period of the electromagnetic wave. If the pulse is continued up to the completion of the period, then the contributions from both ends would almost compensate each other. It can be observed from figure 4.5 that once the alternating electric field is turned off, the quantum flux calculated by the pulse with an odd number of half-cycles is oscillating with a considerably larger amplitude compared to the one for a pulse with an even number of half-cycles.

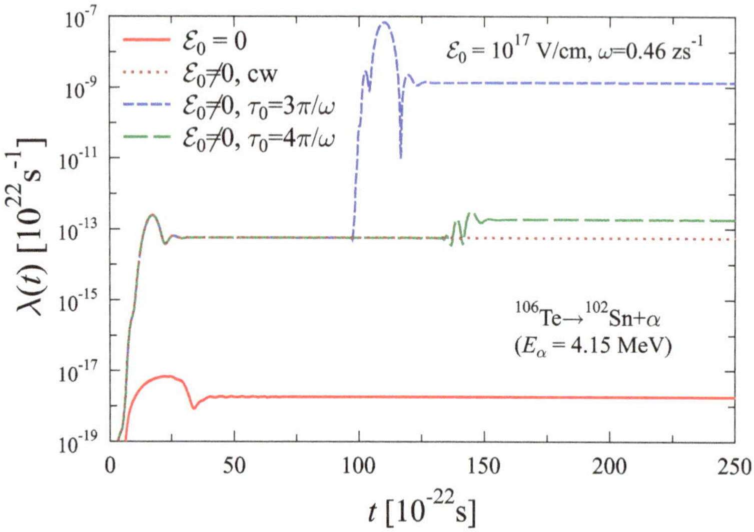

Figure 4.4. Decay rates of the α particle from ^{106}Te for four cases: field-free case (full curve), continuous wave (dotted), short pulse of three half-cycle duration (short dashes), and short pulse of four half-cycle duration (long dashes). The values of the radiation field and frequency are $\mathscr{E}_0 = 10$ keV fm^{-1} and $\omega = 0.46$ zs^{-1}. Note that the repetition of the pulse is made after a time equal to the corresponding pulse duration. Reproduced from [3]. Copyright IOP Publishing Ltd.

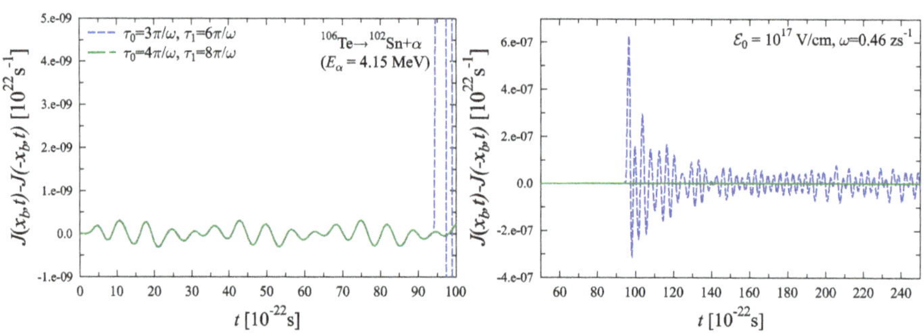

Figure 4.5. Total flux across the nuclear domain in the case of a short pulse of three half-cycle duration (short dashes), and a second one of four half-cycle duration (long dashes). The values of the radiation field and frequency are $\mathscr{E}_0 = 10$ keV fm^{-1} and $\omega = 0.46$ zs^{-1}. Note that the repetition of the pulse is made after a time equal to the corresponding pulse duration. Reproduced from [3]. Copyright IOP Publishing Ltd.

4.2.2 Laser-assisted proton decay in a two-dimensional approach

Like α-decay, proton emission is a spontaneous radioactive process common to several spherical [45–47] and deformed [48–51] proton-rich nuclei. However this last process, in contrast to the former, is characterized by the essential feature that the quasi-bound proton (decay energy $E_p > 0$) is emitted from a state with non-vanishing angular momentum. Up to now the proton emission from nuclei exposed to a strong electromagnetic field was considered in the context of the nuclear photoeffect: concomitantly with the action of an intense laser field, the decaying nucleus absorbs a single γ-ray photon [11, 12]. In this section the proton emission from spherical and deformed nuclei irradiated by strong laser pulses of various photon energies and intensities is considered.

The TDSE describing the center-of-mass motion for a proton ($Z_1 = 1$) and daughter nucleus ($Z_2 = Z - 1$), where Z is the charge number of the mother nucleus, reads in the *length gauge*

$$i\hbar\frac{\partial\psi(\mathbf{r},\, t)}{\partial t} = [H_0 + H_{\text{int}}]\psi(\mathbf{r},\, t), \tag{4.49}$$

where $m = (1 - 1/A)m_p$ is approximately equal to the mass of the proton m_p, and $Z_{\text{eff}} = 1 - Z/A$ is the proton charge for dipole electric transitions [52]. The corresponding unperturbed Hamiltonian is

$$H_0 = -\frac{\hbar^2}{2m}\nabla_r^2 + V(\mathbf{r}), \tag{4.50}$$

whereas the Hamiltonian describing the interaction of the proton-daughter system with the electromagnetic field can be written

$$H_{\text{int}} = -eZ_{\text{eff}}\mathbf{r}\cdot\mathbf{E}(t). \tag{4.51}$$

As shown in chapter 1 and applied to the one-dimensional case in this chapter, the TDSE equation in the *length gauge* can be obtained from the original initial value problem by performing a generalized Göppert-Mayer transformation and operating the dipole approximation on the electromagnetic interaction.

In the previous discussion on α-decay it has been assumed that the direction of the electric field is parallel to the symmetry axis of the dinuclear system. This means that the emission direction of the light particle takes place along the direction of the external field. In the present discussion this severe constraint is removed and consequently the wave-packet of the proton is free to move in three-dimensional space. By considering a linearly polarized laser field, the proton-daughter system displays rotational symmetry about the polarization direction. Owing to the axial symmetry of the problem it is amenable to a two-dimensional description. A convenient system of coordinates is the cylindrical one, because it makes the problem tractable in only two variables: the radial distance ρ and the height z; the azimuthal angle Φ is removed by virtue of the axial symmetry assumption. Equation (4.49) is then rewritten as

$$i\hbar\frac{\partial\psi}{\partial t} = -\frac{\hbar^2}{2m}\left(\frac{1}{\rho}\frac{\partial}{\partial\rho}\rho\frac{\partial}{\partial\rho} + \frac{\partial^2}{\partial z^2} - \frac{\Lambda^2}{\rho^2}\right)\psi + (V(\rho,\, z) + H_{\text{int}})\psi. \tag{4.52}$$

Above $V(\rho, z)$ is the single-particle potential for the quasi-bound proton. The nuclear potential is given by a deformed Woods–Saxon

$$V_{\text{nucl}}(\rho,\, z) = -V_0[1 + \exp(\Theta(\rho,\, z)/a)]^{-1}, \tag{4.53}$$

where V_0 is the potential depth, a the diffuseness, and Θ is an approximation of the distance between an arbitrary space point and the nuclear surface.

The Coulomb potential generated by the charge $(Z - 1)e$, uniformly distributed inside the daughter's nucleus equivalent spherical sharp surface of radius R_0, reads

$$V_{\text{coul}} = \frac{(Z-1)e^2}{4\pi\varepsilon_0}\int \mathrm{d}\mathbf{r}\frac{\rho_c(\mathbf{r}')}{|\mathbf{r}-\mathbf{r}'|}. \tag{4.54}$$

The axially symmetric nuclear shape is described via the Cassini parametrization, i.e. $(\alpha, \alpha_{m=1,2,\ldots})$ (see [53] for details).

The main goal of the present discussion is to analyze the proton decay process in the field-free and free–free cases when the spin–orbit interaction is neglected. The spin–orbit effects could play a role when the magnetic component of the laser field is important. In the present case non-dipole contributions to the Hamiltonian are not important and therefore are not included in the laser–nucleus coupling. In either way, the presence of the spin–orbit term implies a solution with two components, i.e. spin-up and spin-down, that ultimately leads to an increase in computing time without gaining additional insight into the role played by the laser field.

The dipole interaction (4.51) in cylindrical coordinates assumes the form

$$H_{\text{int}}(\rho, z) = -eZ_{\text{eff}}E(t)(z\cos\chi + \rho\sin\chi), \tag{4.55}$$

where χ is the angle between the nucleus symmetry axis, chosen parallel to the z-axis, and the direction of the laser electric field. The electric component of the electromagnetic pulse is assumed to be a modulated monochromatic linearly polarized plane wave,

$$E(t) = \mathscr{E}_0 F(t)\sin\omega t, \tag{4.56}$$

where $\mathscr{E}_0$ is the peak of the pulse and $F(t)$ is the envelope of the pulse.

Note that a wave with circular polarization would induce a coupling between the wavefunction in the angular momentum state Λ with the states with $\Lambda \pm 1$. Such a circumstance introduces a complication since in that case one would deal with a multi-channel problem that must be solved simultaneously in each channel.

In what follows, in a manner similar to the two-dimensional case, numerical experiments using an odd number of half-cycles in the sequence of square pulses are carried out. The task is to validate the conclusion met in the previous section, i.e. the dramatic enhancement of the α-particle tunneling probability when such types of pulses are used.

As already pointed out earlier for α-decay (see also [3]), the dependence of H_{int} on the coordinate z leads to numerical instabilities at the computational boundary. They are mostly related to the linear dependence of the dipole interaction on the coordinate, and in order to remove these numerical instabilities at the computational boundaries, a unitary transformation on the original wavefunction adapted to cylindrical coordinates is performed,

$$\psi(\rho, z) = \mathrm{e}^{-\frac{\mathrm{i}}{\hbar}eZ_{\text{eff}}\int H_{\text{int}}(t')\mathrm{d}t'}\tilde{\psi}(\rho, z). \tag{4.57}$$

Consider first the case of a time-varying electric field (4.56) with pulse shape function $F(t) = 1$. Taking into account that in this case

$$f(z,\, t) \equiv \int H_{\text{int}}(t')\mathrm{d}t' = -\frac{eZ_{\text{eff}}\mathscr{E}_0}{\omega}(z\cos\chi + \rho\sin\chi)(1 - \cos\omega t), \tag{4.58}$$

the TDSE is modified only with respect to the z-component of the momentum,

$$i\hbar\frac{\partial\tilde{\psi}}{\partial t} = -\frac{\hbar^2}{2m}\left(\frac{D}{D\rho}\frac{D}{D\rho} + \frac{1}{\rho}\frac{D}{D\rho} + \frac{D}{Dz}\frac{D}{Dz} + V(\rho,\, z) - \frac{\Lambda^2}{\rho^2}\right)\tilde{\psi}, \tag{4.59}$$

where

$$\begin{Bmatrix} \dfrac{D}{D\rho} \\ \dfrac{D}{Dz} \end{Bmatrix} = \begin{Bmatrix} \dfrac{\partial}{\partial\rho} \\ \dfrac{\partial}{\partial z} \end{Bmatrix} + \frac{ieZ_{\text{eff}}\mathscr{E}_0}{\hbar\omega}(1 - \cos\omega t)\begin{Bmatrix} \sin\chi \\ \cos\chi \end{Bmatrix}. \tag{4.60}$$

It is worth emphasizing that this approach is not restricted to the above simple types of pulse shapes. Other, non-trivial envelopes such as the $\sin^2$ or Gaussian pulses are also tractable in this framework. For example, the pulse with an envelope function

$$F(t) = \sin^2\frac{\pi t}{\tau_p}, \quad \tau_p = N\frac{2\pi}{\omega}, \tag{4.61}$$

will produce, as in the case of the CW pulse (4.58), a simple analytical expression for the $f(t)$ function,

$$\begin{aligned} f(z,\, t) = &-\frac{eZ_{\text{eff}}\mathscr{E}_0}{2\omega}(z\cos\chi + \rho\sin\chi) \\ &\times\left\{1 - \cos\omega t + \frac{N^2}{N^2 - 1}\left[\cos\omega t\cos\left(\frac{\omega t}{N}\right) + \frac{1}{N}\sin\omega t\sin\left(\frac{\omega t}{N}\right)\right]\right\}. \end{aligned} \tag{4.62}$$

In the formula above, the length of the pulse is an entire multiple of the electromagnetic wave period.

The numerical integration of the two-dimensional TDSE is far more demanding in respect to the computer time compared to the one-dimensional case. To speed-up calculations, the two-dimensional treatment of the proton decay from a nucleus under the influence of an ultra-intense time-varying electromagnetic pulse is carried out following [54, 55], where a model to solve the non-stationary problem of spontaneous proton emission from a nucleus in a two-dimensional axial-symmetric geometry was elaborated.

Numerical schemes designed to solve in cylindrical coordinates the two-dimensional single-electron Schrödinger equation are used to calculate, for example, atomic ionization rates in a laser field (see [56] and references therein).

The modified TDSE (4.59) is solved by using the Crank–Nicolson method extended to two spatial dimensions combined with transparent boundary conditions. The scheme is applied by performing a discretization of the spatial grid with the mesh points:

$$\rho_j = \left(j - \frac{1}{2}\right)\Delta\rho, \quad j = 1,\, \dots,\, N_\rho, \quad z_k = k\Delta z, \quad k = -N_z,\, \dots,\, N_z. \tag{4.63}$$

If the wavefunction at the moment $t_n = n\Delta t$ is $\tilde{\psi}^n_{jk}$, the Crank–Nicolson scheme calculates the wavefunction at later times via a linear system. The corresponding matrix has a large dimension ($N_\rho \times (2N_z + 1)$) but fortunately it is sparse. The initial wavefunction at $t = 0$ is obtained by solving the eigenvalue problem corresponding to the two-dimensional Schrödinger equation in a modified potential. By extending the considerations from the one-dimensional case, an equipotential line $\rho_{\text{mod}}(z)$ is defined; it is the most exterior curve, relative to the origin of the potential, on which and beyond, i.e. for $\rho(z) \geqslant \rho_{\text{mod}}(z)$, $V(\rho, z)$ has a constant value slightly larger than the proton decay energy E_p. Accordingly, the initial wavefunction is a bound state in the modified potential and the two-dimensional equation is solved by performing the above discussed discretization procedure in space. The depth of the Woods–Saxon potential is varied until one of the eigenvalues matches the experimental proton decay energy E_p. As already done in the one-dimensional case, to avoid reflections of the propagated wave and therefore errors in the calculations of physical quantities at the grid frontiers, the TBC algorithm is employed at the boundaries. The total physical volume is separated in the interior nuclear volume and an exterior domain, i.e. $\mathscr{D} = \mathscr{D}_{\text{int}} \bigcup \mathscr{D}_{\text{ext}}$, that are delimited by the equipotential line that is defined as the most exterior curve on which $V(\rho, z) = E_p$ (see figure 4.6). The tunneling dynamics is eventually solved on a spatial region of reasonably large dimension, i.e. $z \in [-256\ \text{fm}, 256\ \text{fm}]$ and $\rho \in (0, 256\ \text{fm})$.

In what follows the observables related to the laser-assisted proton radioactivity are calculated.

Any solution of the TDSE satisfies the continuity equation for the density of probability

$$\frac{\partial n}{\partial t} + \nabla \cdot \mathbf{J} = 0, \tag{4.64}$$

with

$$n = |\psi(\rho, z, t)|^2, \tag{4.65}$$

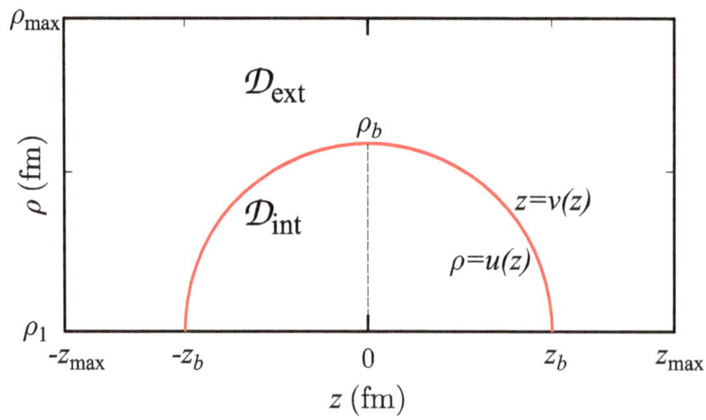

Figure 4.6. A representation of the regions implied in the calculation of the decay rate. Reproduced from [9]. Copyright IOP Publishing Ltd.

$$\mathbf{J}(\rho, z, t) = -\frac{i\hbar}{2\mu}[\psi^*(\rho, z, t)\nabla\psi(\rho, z, t) - \nabla\psi^*(\rho, z, t)\psi(\rho, z, t)]. \tag{4.66}$$

Using the transformation given by equation (4.57) and the form of the operator ∇ in cylindrical coordinates, equations (4.65) and (4.66) are rewritten as

$$n = |\tilde{\psi}(\rho, z, t)|^2, \tag{4.67}$$

$$\begin{aligned}\mathbf{J}(\rho, z, t) = &-\frac{i\hbar}{2\mu}[\tilde{\psi}^*(\rho, z, t)\nabla\tilde{\psi}(\rho, z, t) - \tilde{\psi}(\rho, z, t)\nabla\tilde{\psi}^*(\rho, z, t)]\\ &-\frac{2i}{\hbar}\mathbf{e}_z\frac{\partial f(z, t)}{\partial z}|\tilde{\psi}(\rho, z, t)|^2.\end{aligned} \tag{4.68}$$

Note that only the z-component of the current operator is modified.

The total proton tunneling probability is estimated outside the nuclear volume

$$P_{\text{tun}} = 2\pi\int_{\mathscr{D}_{\text{ext}}} n(\rho, z, t)\rho d\rho dz, \tag{4.69}$$

and the proton total decay rate is expressed as

$$\lambda(t) = \frac{1}{N_{\text{tot}} - P_{\text{tun}}}\frac{dP_{\text{tun}}}{dt}, \tag{4.70}$$

where N_{tot} is the total norm given by

$$N_{\text{tot}} = 2\pi\int_{\mathscr{D}} n(\rho, z, t)\rho d\rho dz. \tag{4.71}$$

The calculation of dP_{tun}/dt requires at first sight the evaluation of a two-dimensional integral on the exterior domain $\mathscr{D}_{\text{ext}}$. However, by using the continuity equation (4.64) and a few rules from the integral calculus, a major simplification occurs such that eventually one-dimensional integrals are left to be evaluated [54].

The procedure can be carried out by expressing first the flux $\nabla \cdot \mathbf{J}$ in cylindrical coordinates,

$$\nabla \cdot \mathbf{J} = -\frac{i\hbar}{2\mu}\left\{\frac{1}{\rho}\frac{\partial}{\partial\rho}\left[\rho\left(\psi^*\frac{\partial\psi}{\partial\rho} - \psi\frac{\partial\psi^*}{\partial\rho}\right)\right] + \frac{\partial}{\partial z}\left(\psi^*\frac{\partial\psi}{\partial z} - \psi\frac{\partial\psi^*}{\partial z}\right)\right\}, \tag{4.72}$$

and then, using the continuity equation (4.64) to express the time derivative of the tunneling probability,

$$\begin{aligned}\frac{dP_{\text{tun}}(t)}{dt} &= \frac{d}{dt}\left(2\pi\int_{\mathscr{D}_{\text{ext}}}|\psi(\rho, z, t)|^2\rho d\rho dz\right) = -2\pi\int_{\mathscr{D}_{\text{ext}}}\nabla \cdot \mathbf{J}\rho d\rho dz\\ &= 2\pi\frac{i\hbar}{2\mu}\left\{\int_{\mathscr{D}_{\text{ext}}}\frac{\partial}{\partial\rho}\left[\rho\left(\psi^*\frac{\partial\psi}{\partial\rho} - \psi\frac{\partial\psi^*}{\partial\rho}\right)\right]d\rho dz\right.\end{aligned} \tag{4.73}$$

$$+\int_{\mathscr{D}_{\rm ext}} \frac{\partial}{\partial z}\left[\rho\left(\psi^*\frac{\partial\psi}{\partial z} - \psi\frac{\partial\psi^*}{\partial z}\right)\right]{\rm d}\rho{\rm d}z\Bigg\}, \tag{4.74}$$

one can eventually write

$$\frac{{\rm d}P_{\rm tun}(t)}{{\rm d}t} = 2\pi\frac{{\rm i}\hbar}{2\mu}\left(\int_{\mathscr{D}_{\rm ext}} \frac{\partial\mathscr{P}}{\partial\rho}{\rm d}\rho{\rm d}z + \int_{\mathscr{D}_{\rm ext}} \frac{\partial\mathscr{Q}}{\partial z}{\rm d}\rho{\rm d}z\right) = 2\pi\frac{{\rm i}\hbar}{2\mu}(I_1 + I_2). \tag{4.75}$$

The following notations have been introduced above:

$$\mathscr{P}(\rho, z) = \rho\left(\psi^*\frac{\partial\psi}{\partial\rho} - \psi\frac{\partial\psi^*}{\partial\rho}\right),\quad \mathscr{Q}(\rho, z) = \rho\left(\psi^*\frac{\partial\psi}{\partial z} - \psi\frac{\partial\psi^*}{\partial z}\right). \tag{4.76}$$

The external domain, $\mathscr{D}_{\rm ext}$, is delimited by the above defined equipotential and the limits of the numerical domain. The equipotential line can be expressed either as a function of z, $\rho = u(z)$, or as a function of ρ, $z = v(\rho)$. In figure 4.6 the represented curve corresponds to a spherical case, but the same rationale can be made for a deformed shape. Applying the theorem of Green, the double integrals are transformed into curvilinear integrals. Using the notations introduced in figure 4.6, the first integral appearing in the rhs of (4.75) can be presented in the form

$$\begin{aligned} I_1 &= \int_{-z_{\max}}^{-z_b} {\rm d}z \int_{\rho_1}^{\rho_{\max}} \frac{\partial\mathscr{P}}{\partial\rho}{\rm d}\rho + \int_{-z_b}^{z_b} {\rm d}z \int_{\rho=u(z)}^{\rho_{\max}} \frac{\partial\mathscr{P}}{\partial\rho}{\rm d}\rho + \int_{z_b}^{z_{\max}} {\rm d}z \int_{\rho_1}^{\rho_{\max}} \frac{\partial\mathscr{P}}{\partial\rho}{\rm d}\rho \\ &= \int_{-z_{\max}}^{-z_b} \left[\mathscr{P}(\rho_{\max}, z) - \mathscr{P}(\rho_1, z)\right]{\rm d}z + \int_{-z_b}^{z_b} \left[\mathscr{P}(\rho_{\max}, z) - \mathscr{P}(u(z), z)\right]{\rm d}z \\ &\quad + \int_{z_b}^{z_{\max}} \left[\mathscr{P}(\rho_{\max}, z) - \mathscr{P}(\rho_1, z)\right]{\rm d}z. \end{aligned} \tag{4.77}$$

In this last formula, $-z_b$ and z_b are the points where the equipotential curve crosses the axis $\rho = \rho_1$. The considered cases have a symmetry with respect to $z = 0$, so that we can limit ourselves to a half of domain. When there is no such a symmetry, the above formula can be used with an appropriate adaptation.

Thus, for the configuration illustrated in figure 4.6, I_1 reads:

$$\frac{1}{2}I_1 = \int_0^{z_b} \left[\mathscr{P}(\rho_{\max}, z) - \mathscr{P}(u(z), z)\right]{\rm d}z + \int_{z_b}^{z_{\max}} \left[\mathscr{P}(\rho_{\max}, z) - \mathscr{P}(\rho_1, z)\right]{\rm d}z.$$

The second integral appearing in (4.75) can be obtained in a similar manner,

$$\begin{aligned} \frac{1}{2}I_2 &= \int_{\rho_1}^{\rho_b} {\rm d}\rho \int_{z=v(\rho)}^{z_{\max}} \frac{\partial\mathscr{Q}}{\partial z}{\rm d}z + \int_{\rho_b}^{\rho_{\max}} {\rm d}\rho \int_0^{z_{\max}} \frac{\partial\mathscr{Q}}{\partial z}{\rm d}z \\ &= \int_{\rho_1}^{\rho_b} [\mathscr{Q}(\rho, z_{\max}) - \mathscr{Q}(\rho, v(\rho))]{\rm d}\rho + \int_{\rho_b}^{\rho_{\max}} [\mathscr{Q}(\rho, z_{\max}) - \mathscr{Q}(\rho, 0)]{\rm d}\rho. \end{aligned}$$

Note that the equipotential curve meets the axis $z = 0$ at the point specified by ρ_b, and that terms containing the values $\rho_{\max}$ and $z_{\max}$, belonging to the boundaries, are considered zero because when $\rho_{\max}, z_{\max} \to \infty$, the wavefunction tends asymptotically to zero. Therefore,

$$\frac{1}{2}I_1 = -\int_0^{z_b} \mathscr{P}(u(z), z)\mathrm{d}z - \int_{z_b}^{z_{\max}} \mathscr{P}(\rho_1, z)\mathrm{d}z, \tag{4.78}$$

and

$$\frac{1}{2}I_2 = -\int_{\rho_1}^{\rho_b} \mathscr{Q}(\rho, v(\rho))\mathrm{d}\rho - \int_{\rho_b}^{\rho_{\max}} \mathscr{Q}(\rho, 0)\mathrm{d}\rho. \tag{4.79}$$

When performing the numerical calculations, $\psi(\rho_1, z)$ for $z > z_{\max}$, and $\psi(\rho, 0)$ for $\rho > \rho_{\max}$, are neglected, since they are usually small. Using the notations

$$\psi = a + \mathrm{i}b, \quad \frac{\partial\psi}{\partial\rho} = a_\rho + \mathrm{i}b_\rho, \quad \frac{\partial\psi}{\partial z} = a_z + \mathrm{i}b_z,$$

the functions defined in (4.76) are rewritten as

$$\mathscr{P}(\rho, z) = 2\mathrm{i}\rho(ab_\rho - a_\rho b), \quad \mathscr{Q}(\rho, z) = 2\mathrm{i}\rho(ab_z - a_z b).$$

Using equations (4.78) and (4.79), I_1, I_2 and subsequently $\mathrm{d}P_{\mathrm{tun}}(t)/\mathrm{d}t$ from equation (4.75) can be obtained. Note that since in practice the wavefunction values are known only at the grid points, the equipotential line, the first derivatives with respect to ρ and z, as well the one-dimensional integrals, are evaluated approximately.

4.2.3 Some cases of proton decay in spherical and deformed nuclei

The calculations for the decay rate (4.70) are exemplified below by specializing to the case when the electric field is parallel to the nucleus symmetry axis ($\chi = 0$). Various laser intensities and frequencies for either continuous waves, when the electromagnetic field acts from $t = 0$ to $t = \infty$, or few-cycle pulses are considered. The evolution in time of the decay rate is followed on an interval with step $\Delta t = 2.5 \times 10^{-23}$ s. Three cases of proton decay involving one spherical and two deformed mother nuclei are considered. The depth of the nuclear potential V_0 is adjusted in order to reproduce the experimental value E_p.

4.2.3.1 $^{171}\mathrm{Au} \to {}^{170}\mathrm{Pt} + \mathrm{p}$

The proton radioactivity of ^{171}Au has been reported by C N Davis and collaborators at Argonne [57]. The proton emission from the $j^\pi = 11/2^-$ state ($E_p = 1.718$ MeV) of the spherical nucleus ^{171}Au, as taken in the compilation [58], is analyzed below. Hence $\alpha = 0$ in the Cassini parametrization and the angular momentum appearing in the TDSE (4.52) is $\Lambda = 5$. The total potential is shown in figure 4.7 in a three-dimensional representation.

The evolution in time of the proton decay rate of ^{171}Au is represented in the left panel of figure 4.8 in the field-free case ($\mathscr{E}_0 = 0$) and in the laser-assisted case ($\mathscr{E}_0 \neq 0$)

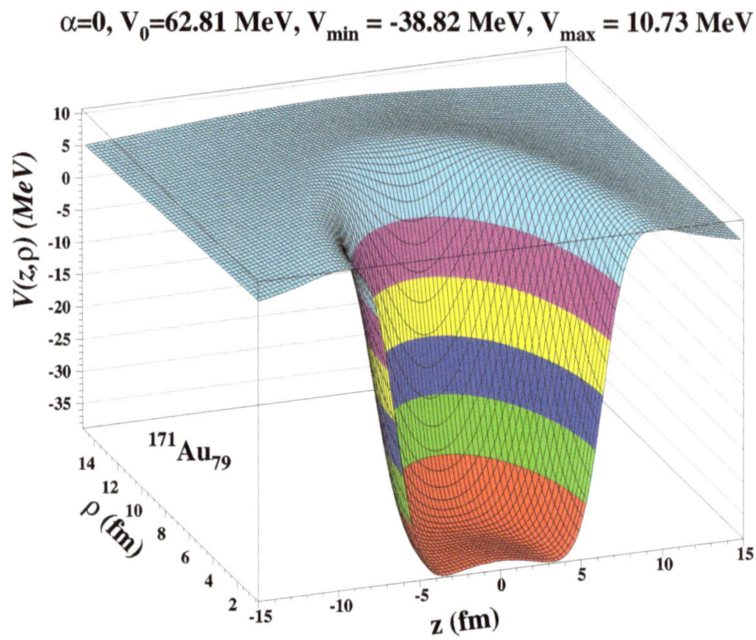

Figure 4.7. Three-dimensional plot of the total potential for the spherical nucleus ^{171}Au. By V_0 we denote the Woods–Saxon potential depth that fits the decay energy E_p measured for this nucleus, $V_{\min}$ is the potential minimum and $V_{\max}$ is the barrier height. Reproduced from [9]. Copyright IOP Publishing Ltd.

for a radiation frequency chosen at the upper limit of the x-ray spectrum ($\hbar\omega = 100$ keV) and an electric field strength $\mathscr{E}_0 = 10^{10}$ V cm^{-1} ($e\mathscr{E}_0 = 10^{-9}$ MeV fm^{-1}). This strength corresponds to the relative moderate intensity $I_0 = 1.32 \times 10^{17}$ W cm^{-2}. In the right panel of figure 4.8 the same quantity is analyzed in the case of a Ti:sapphire laser ($\hbar\omega = 1.5$ eV) and for the much higher intensity $I_0 = 1.32 \times 10^{25}$ W cm^{-2} ($\mathscr{E}_0 = 10^{14}$ V cm^{-1}). It should be noted that for both x-ray and optical lasers, the decay rate $\lambda(t)$ increases rapidly to a maximum value after a transient time and, subsequently, oscillates with the same period but with different amplitudes around an asymptotic value, which coincides with the asymptotic value in the exponential regime of the decay [54]. In the second case, although the radiation frequency is $\sim 10^5$ smaller than in the first case, the intensity is $\sim 10^8$ larger and the induced oscillations in the decay rate are more pronounced. For both cases the permanent action of the laser field, represented by the dashed lines in figure 4.8, does not change in the overall asymptotic decay rate $\lambda(\infty)$.

The occurrence of a high-frequency oscillation of the decay rate in the laser-assisted case, with the same period independent of the external field frequency and intensity, as illustrated by the insets in figure 4.8, deserves a few comments. When preparing the initial bound wavefunction, a second eigenstate is obtained, which lays at 10.7 MeV in the case of ^{171}Au. The difference between the second $|2\rangle$ and the first $|1\rangle$ eigenstate is approximately 9 MeV. Consequently at early times there are two quasi-bound levels inside the potential well that substantially retain the character of the initially bound states: a low-lying one, from where proton emission occurs and a

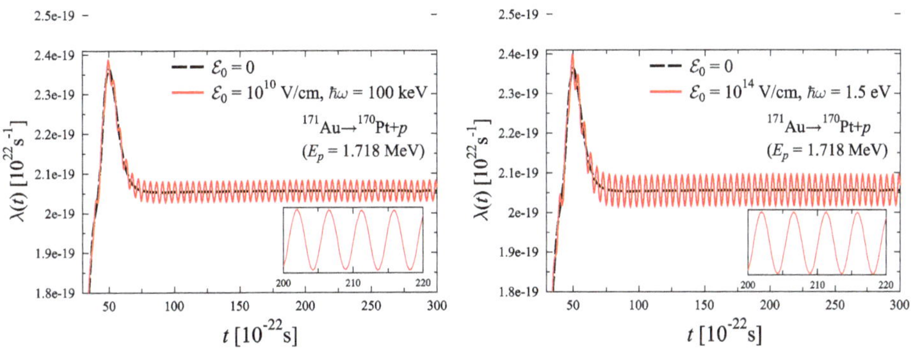

Figure 4.8. Decay rates (4.70) for the proton decay rate of ^{171}Au in the field-free case (black curve) assisted by a CW laser pulse of amplitude $\mathcal{E}_0 = 10^{10}$ V cm^{-1} ($I_0 = 1.32 \times 10^{17}$ W cm^{-2}), and photon energy $\hbar\omega = 100$ keV (left panel) and $\mathcal{E}_0 = 10^{14}$ V cm^{-1} ($I_0 = 1.32 \times 10^{25}$ W cm^{-2}), $\hbar\omega = 1.5$ eV (right panel). The magnification of the laser-assisted decay rates on the time interval $t \in [20zs, 30zs]$ is represented in the inserts. Reproduced from [9]. Copyright IOP Publishing Ltd.

high-lying state that initially has zero occupation probability, and that once the external field is turned on, starts to be occupied with a probability given by [59]:

$$P_{1\to 2}(t) \sim \sin^2 \frac{\Omega t}{2}, \tag{4.80}$$

where $\Omega = \Delta E_{12}/\hbar$. The above dependence was obtained under the assumption that the transition matrix element of the interaction is much smaller than the spacing between the two states ΔE_{12},

$$eZ_{\text{eff}}\mathcal{E}_0|\langle 1|z|2\rangle| \ll \Delta E_{12}.$$

It can be shown that for ^{171}Au, $\langle 1|z|2\rangle \approx 1.52$ and choosing the case with the highest amplitude of the electric field, $\mathcal{E}_0 = 10^{14}$ V cm^{-1}, the lhs of (4.2.3.1) is $\approx 8.2 \times 10^{-6}$ MeV. Figure 4.8 shows that on the curve $\mathrm{d}P_{\text{tun}}/\mathrm{d}t$, corresponding to $\mathcal{E}_0 = 0$, a smaller signal $\mathrm{d}P_{1\to 2}/\mathrm{d}t$ is superposed in the laser-assisted case. The decay rate is then weakly perturbed and exhibits small oscillations around the asymptotic value as dictated by the periodic function $\sin \Omega t$. This conjecture is confirmed by the fact that the period of oscillations, read off by magnifying figure 4.8, is $T_{\text{osc}} \approx 4.5 \times 10^{-22}$ s, a value which is quite close to the double of the oscillation period of the transition probability (4.80), i.e. $2\pi\hbar/\Delta E_{12} \approx 4.6 \times 10^{-22}$ s. This behavior is altered at later times since the two states are departing from the type described by bound wavefunctions.

It was shown in the previous section that a spectacular enhancement of the α-decay rate takes place in the one-dimensional case when the external electromagnetic field acts in the idealized form of a short pulse with a duration of an odd number of half-cycles. In the present two-dimensional approach this numerical experiment is repeated for the case when $\mathcal{E}_0 = 10^{10}$ V cm^{-1} and $\hbar\omega = 100$ keV: a jump of the ^{171}Au proton decay rate of almost three orders of magnitude results, as can be inferred from figure 4.9. Since the external field is switched-off at time

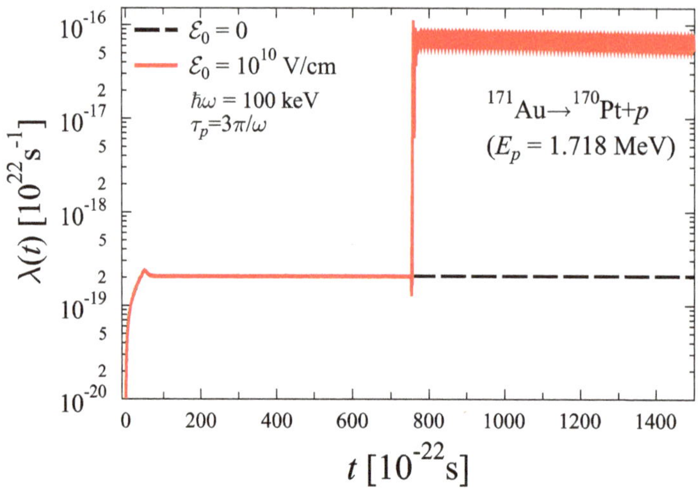

Figure 4.9. Decay rates (4.70) for the proton decay rate of ^{171}Au in the field-free case (dashed curve) and assisted by a laser pulse of duration $\tau_p = 3\pi/\omega$ and amplitude $\mathcal{E}_0 = 10^{10}$ V cm^{-1} and photon energy $\hbar\omega = 100$ keV. Reproduced from [9]. Copyright IOP Publishing Ltd.

$t = 3T/2$, where $T = 2\pi/\omega$ is the period of the electromagnetic wave, $\lambda(t)$ for $\mathcal{E}_0 \neq 0$ decrease exponentially to the asymptotic value in the field-free case. Note that in the above analysis, for practical purposes, the light pulse was turned on at $t = 0$. If a sequence of such pulses with an odd number of half-cycles is applied to the decaying system, the net effect on the asymptotic value of the decay rate does not depend on the time origin. After the completion of the light pulse sequence, the decay rate falls back smoothly to the field-free asymptotic value.

4.2.3.2 $^{141}Ho \rightarrow {}^{140}Dy + p$

Proton radioactivity from ^{141}Ho was reported by C N Davids *et al* following measurements using the Argonne Fragment Mass Analyzer (FMA) [60]. Subsequent measurements of the level structure of the deformed proton emitter ^{141}Ho by means of in-beam γ-ray spectroscopy and particle-rotor calculations of the energy levels in the rotational band lying above the $I = 7/2^-$ ground state suggest that a better agreement with experiment would be obtained if the nuclear shape is assumed to exhibit hexadecapole deformation and triaxiality [51, 61].

The proton emission of the nucleus ^{141}Ho from a state with angular momentum $\Lambda = 2$ and energy $E_p = 1.169$ MeV is analyzed in the case when significant quadrupole and hexadecupole deformations are present, i.e. $\beta_2 = 0.31$, $\beta_4 = -0.04$ [58]. In the Cassini parametrization of the deformed surface this corresponds approximately to $\alpha = 0.27$ (see figure 4.10).

The proton decay rate of ^{141}Ho is represented in the left panel of figure 4.11 in the field-free case and in the laser-assisted case for a x-ray photon ($\hbar\omega = 100$ keV) and an electric field strength $\mathcal{E}_0 = 10^{10}$ V cm^{-1}. In the right panel of figure 4.11 the same quantity is analyzed for the optical laser ($\hbar\omega = 1.5$ eV) and electric field strength

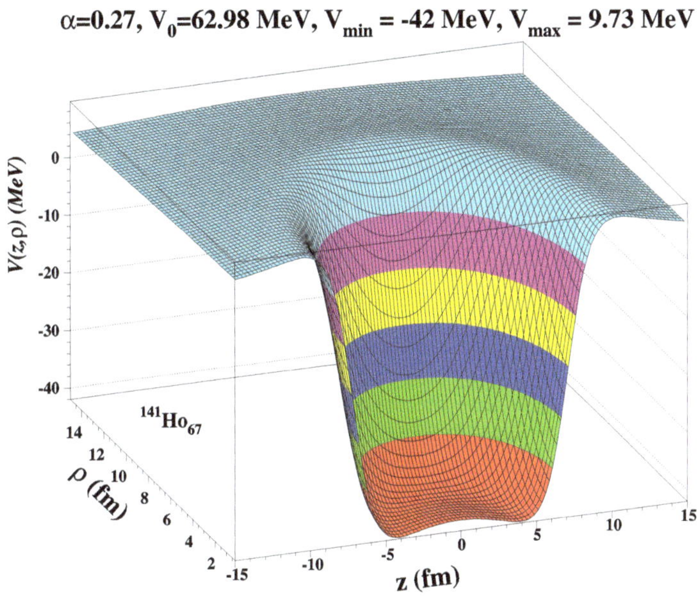

Figure 4.10. Three-dimensional plot of the total potential for the deformed nucleus ^{141}Ho. Reproduced from [9]. Copyright IOP Publishing Ltd.

$\mathscr{E}_0 = 10^{15}$ V cm^{-1}. Similarly to the previous case, in the presence of the laser field, $\lambda(t)$ oscillates around the field-free value. Like in the previous case, the laser field do not change in overall the asymptotic decay rate $\lambda(\infty)$. Although the case with an optical laser has an intensity $\sim 10^{10}$ larger than the x-ray laser, the amplitude of these oscillations are rather close. Due to the lower mass and high deformation of ^{141}Ho the decay rates are higher than those for ^{171}Au despite the lower decay energy in the former case.

In the presence of a short laser pulse with duration $3T/2$ and characteristics $\mathscr{E}_0 = 10^{10}$ V cm^{-1}, $\hbar\omega = 100$ keV, the proton decay rate of ^{141}Ho displays a step-like increase once the pulse is turned off, as in the case of ^{171}Au (see figure 4.12). However, in this case the jump is only two orders of magnitude compared to the field-free value. The rapid exponential decay of $\lambda(t)$ to the field-free asymptotic value in the regime $t > \tau_p$ is more clearly put in evidence compared to ^{171}Au, since the proton tunnels through a thinner barrier for ^{141}Ho.

4.2.3.3 $^{145}\text{Tm} \rightarrow {}^{144}\text{Er} + p$

For nuclei around Tm it is expected that the orbitals $s_{1/2}$, $d_{3/2}$, and $h_{11/2}$ dominate the structure of the low-lying states [62]. In the particular case of ^{145}Tm the ground state seems to be dominated by the $h_{11/2}$ single-particle orbital and can decay by proton emission [63]. The study of ^{145}Tm proton radioactivity has revealed that one of these states decays predominantly by the emission of the $h_{11/2}$ proton to the 0^+ ground state of ^{144}Er. The particle–vibration coupling model confirmed this scenario [64].

The proton emission of the nucleus ^{145}Tm is assumed to take place from the state of angular momentum $\Lambda = 5$ and energy $E_p = 1.728$ MeV. The compilation [58]

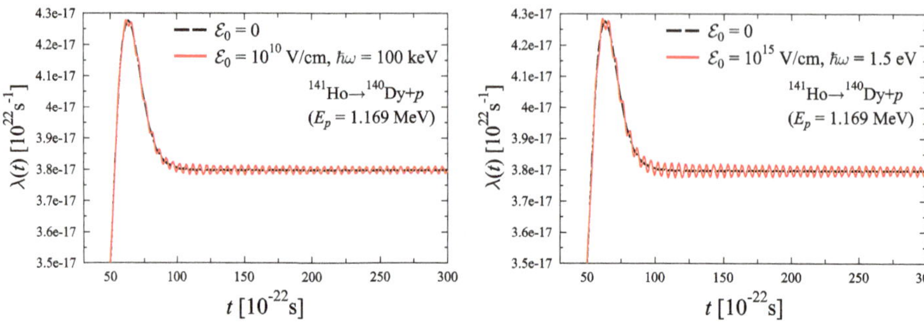

Figure 4.11. Decay rate (4.70) for the proton decay rate of ^{141}Ho in the field-free case (dashed curve) and assisted by a CW laser pulse of amplitude $\mathcal{E}_0 = 10^{10}$ V cm^{-1} ($I_0 = 1.32 \times 10^{17}$ W cm^{-2}) and photon energy $\hbar\omega = 100$ keV (left panel) and for the Ti:sapphire ($\hbar\omega = 1.5$ eV) with $\mathcal{E}_0 = 10^{15}$ V cm^{-1} ($I_0 = 1.32 \times 10^{27}$ W cm^{-2}) (right panel). Reproduced from [9]. Copyright IOP Publishing Ltd.

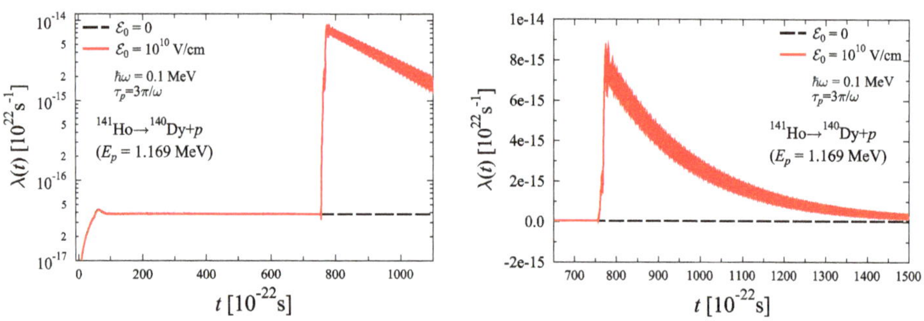

Figure 4.12. Decay rate (4.70) for the proton decay rate of ^{141}Ho in the field-free case (dashed curve) and assisted by a laser pulse of duration $\tau_p = 3\pi/\omega$ and of amplitude $\mathcal{E}_0 = 10^{10}$ V cm^{-1} ($I_0 = 1.32 \times 10^{17}$ Wcm^{-2}), and photon energy $\hbar\omega = 100$ keV. The left panel displays the decay rate in logarithmic scale in order to substantiate the jump at $t = 3\pi/\omega$ compared to the field-free case. The right panel shows the exponential decrease of $\lambda(t)$ after the aforementioned jump. Reproduced from [9]. Copyright IOP Publishing Ltd.

provides non-vanishing quadrupole and hexadecupole deformations: $\beta_2 = 0.26$, $\beta_4 = -0.04$, that are approximated in the Cassini parametrization by the deformation parameter $\alpha = 0.225$ (see figure 4.13). One should note that calculations with the particle-rotor model and the core quasiparticle coupling model indicate that the properties of the $h_{11/2}$ band and the proton decay rates in ^{145}Tm are consistent with the presence of triaxiality [65].

The time evolution of the proton decay rate of ^{145}Tm is represented in the left panel of figure 4.14 in the field-free case and assisted by an x-ray laser of frequency $\hbar\omega = 100$ keV of electric field strength $\mathcal{E}_0 = 10^{11}$ V cm^{-1} ($e\mathcal{E}_0 = 10^{-8}$ MeV fm^{-1}) which is ten times larger than the strengths used for the irradiation of ^{171}Au and ^{141}Ho. This larger value determines stronger oscillations of $\lambda(t)$ around the field-free value. This curve is compared in the right panel of figure 4.14 to the optical laser-assisted corresponding curve, when $\hbar\omega = 1.5$ eV and $\mathcal{E}_0 = 10^{15}$ V cm^{-1}.

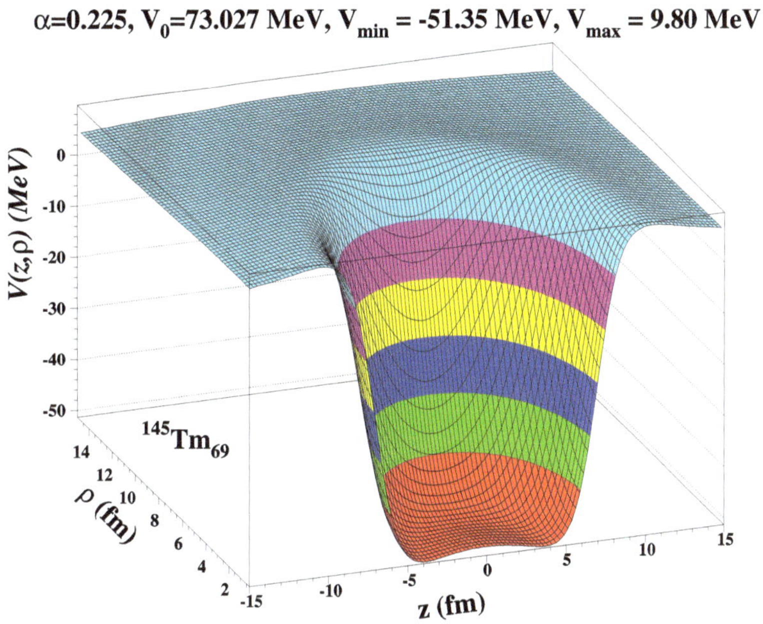

Figure 4.13. Three-dimensional plot of the total potential for the deformed nucleus ^{145}Tm. Reproduced from [9]. Copyright IOP Publishing Ltd.

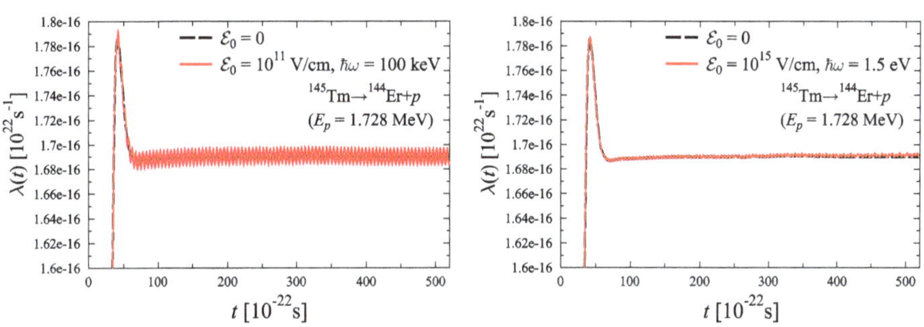

Figure 4.14. Decay rate (4.70) for the proton decay rate of ^{145}Tm in the field-free case (dashed curve) and assisted by a CW laser pulse of amplitude $\mathcal{E}_0 = 10^{11}$ V cm^{-1} ($I_0 = 1.32 \times 10^{19}$ W cm^{-2}) and photon energy $\hbar\omega = 100$ keV (left panel) and $\mathcal{E}_0 = 10^{15}$ V cm^{-1} ($I_0 = 1.32 \times 10^{27}$ W cm^{-2}), $\hbar\omega = 1.5$ eV. Reproduced from [9]. Copyright IOP Publishing Ltd.

As can be inferred from figure 4.15, if the x-ray laser pulse is turned off after three half-cycles, $\lambda(t)$ displays a jump which is smaller than in the previously examined two cases. The oscillations in the transitory regime, leading to the asymptotic field-free value, are instead more pronounced, due to both the higher laser intensity and the higher proton decay energy of ^{145}Tm. The proton encounters the lowest barrier from all three cases and therefore $\lambda(t)$ increases to the a higher maximum compared to the other two cases.

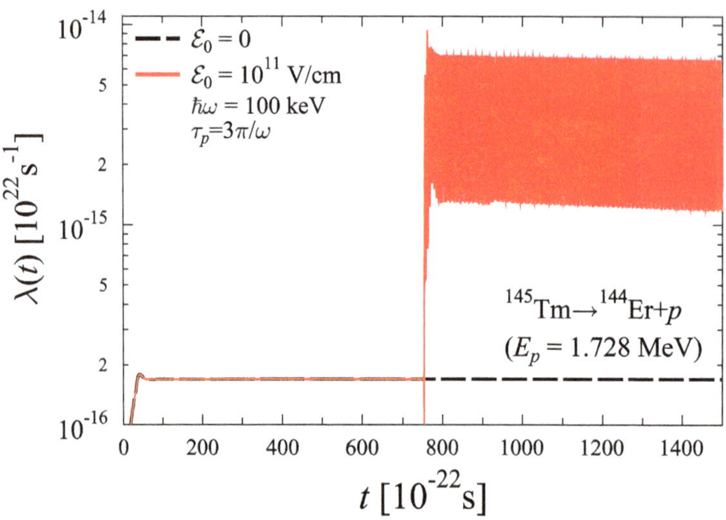

Figure 4.15. Decay rate (4.70) for the proton decay rate of ^{145}Tm in the field-free case (dashed curve) and assisted by a laser pulse of duration $\tau_p = 3\pi/\omega$ with $\mathscr{E}_0 = 10^{11}$ V cm^{-1} ($I_0 = 1.32 \times 10^{19}$ W cm^{-2}) and $\hbar\omega = 100$ keV. Reproduced from [9]. Copyright IOP Publishing Ltd.

References

[1] Di Piazza A, Müller C, Hatsagortsyan K Z and Keitel C H 2012 Extremely high-intensity laser interactions with fundamental quantum systems *Rev. Mod. Phys.* **84** 1177–228

[2] Castañeda Cortés H M 2012 Laser assisted α-decay *Dissertation* 723 (Heidelberg: University of Heidelberg) p 135
Castañeda Cortés H M, Müller M, Keitel C H and Palffy A 2013 Nuclear recollisions in laser-assisted α decay *Phys. Lett.* B **273** 401–5

[3] Mişicu Ş and Rizea M 2013 α-decay in ultra-intense laser fields *J. Phys. G: Nucl. Part. Phys.* **40** 095101

[4] Mişicu Ş and Rizea M 2016 Speeding of α decay in strong laser fields *Open Phys.* **14** 81–7

[5] Kopytin I V and Kornev A S 2014 Semiclassical formula for the alpha-decay rate in an electromagnetic field *Phys. At. Nucl.* **77** 53–9

[6] Delion D S and Ghinescu S A 2017 Geiger-Nuttall law for nuclei in strong electromagnetic fields *Phys. Rev. Lett.* **119** 202501

[7] Kis D P and Szilvasi R 2018 Three dimensional α-tunneling in intense laser fields *J. Phys. G: Nucl. Part. Phys.* **45** 045103

[8] Bai D, Deng D and Ren Z 2018 Charged particle emissions in high-frequency alternative electric fields *Nucl. Phys.* A **976** 23–32

[9] Mişicu Ş and Rizea M 2019 Laser-assisted proton radioactivity of spherical and deformed nuclei *J. Phys. G: Nucl. Part. Phys.* **46** 115106

[10] Cheng J-H, Li Y and Yu T-P 2022 Systematic study of laser-assisted proton radioactivity from deformed nuclei *Phys. Rev.* C **105** 024312

[11] Dadi A and Müller C 2012 Laser-assisted nuclear photoeffect *Phys. Rev.* C **85** 064604
Dadi A 2012 Nuclear processes in intense light–matter interaction *Dissertation* University of Heidelberg p 96

[12] Kalman P, Kis D and Keszthelyi T 2013 Near-threshold laser-modified emission in the nuclear photoeffect *Phys. Rev.* A **87** 063415
[13] Büttiker M and Landauer R 1982 Traversal time for tunneling *Phys. Rev. Lett.* **49** 1739–42
[14] Ivlev B I and Mel'nikov V I 1985 Stimulation of tunneling by a high-frequency field: decay of the zero-voltage state in Josephson junctions *Phys. Rev. Lett.* **55** 1614–7
[15] Onuki A 1994 Quantum nucleation in helium under high-frequency field *J. Phys. Soc. Jpn* **63** 1245–9
[16] Tanizawa T 1996 Quantum tunneling rate in oscillating fields *J. Phys. Soc. Jpn* **65** 3157–66
[17] Tanizawa T 1998 Quantum tunneling rate through an inverted parabola with an oscillating external field *J. Phys. Soc. Jpn* **67** 699–815
[18] Baz' A I, Zeldovich Ya B and Perelomov A M 1971 *Rasseyanie, Reaktsii i Raspady v Nerelyativistskoi Kvantovoi Mehanike (Scattering, Reactions and Decay in Nonrelativistic Quantum Mechanics)* 2nd edn (Moscow: Nauka)
[19] Schwoerer E H, Magill J and Beleites B 2006 *Lasers and Nuclei* (Berlin: Springer)
[20] Emery G T 1972 Perturbation of nuclear decay rates *Ann. Rev. Nucl. Sci.* **22** 165–202
[21] Kettner K U, Becker H W, Strieder F and Rolfs C 2006 High-Z electron screening: the cases ^{50}V(p,n)^{50}Cr and ^{176}Lu(p,n)^{176}Hf *J. Phys. G: Nucl. Part. Phys.* **32** 489–95
[22] Zinner N T 2007 Alpha decay rate enhancement in metals: an unlikely scenario *Nucl. Phys.* A **781** 81–7
[23] Brabec T and Krausz F 2000 Intense few-cycle laser fields: frontiers of nonlinear optics *Rev. Mod. Phys.* **72** 545–91
[24] Seweryniak D *et al* 2006 α decay of ^{105}Te *Phys. Rev.* C **73** 061301(R)
[25] Mohr P 2007 Super-allowed α decay above doubly-magic ^{100}Sn and properties of $^{104}\mathrm{Te} = {}^{100}\mathrm{Sn} \otimes \alpha$ *Eur. Phys. J.* A **31** 23–8
[26] Grossman F 2008 *Theoretical Femtosecond Physics: Atoms and Molecules in Strong Laser Fields* (Berlin: Springer)
[27] Pauli W 1973 *Electrodynamics* (Cambridge, MA: MIT Press)
[28] Chou C-C and Wyatt R E 2013 Time-dependent Schrödinger equation with Markovian outgoing wave boundary conditions: applications to quantum tunneling dynamics and photoionization *Int. J. Quantum Chem.* **113** 39–44
[29] Crank J and Nicolson P 1947 A practical method for numerical evaluation of solutions of partial differential equations of the heat-conduction type *Math. Proc. Camb. Philos. Soc.* **43** 50–67
[30] Askar A and Cakmak A S 1978 Explicit integration method for the time-dependent Schrödinger equation for collision problems *J. Comput. Phys.* **68** 2794–8
[31] Moyer C A 2004 Numerov extension of transparent boundary conditions for the Schrödinger equation in one dimension *Am. J. Phys.* **72** 351–8
[32] Rizea M 2010 Exponential fitting method for the time-dependent Schrödinger equation *J. Math. Chem.* **48** 55–65
[33] van Dijk W and Toyama F M 2007 Accurate numerical solutions of the time-dependent Schrödinger equation *Phys. Rev.* E **75** 036707
[34] Puzynin I V, Selin A V and Vinitsky S I 1999 A high-order accuracy method for numerical solving of the time-dependent Schrödinger equation *Comput. Phys. Commun.* **123** 1–6
[35] Rizea M 2008 On the numerical solution of the time-dependent Schrödinger equation with time-dependent potentials *AIP Conf. Proc.* **1048** 1011–5

[36] Hadley G R 1991 Transparent boundary condition for beam propagation *Opt. Lett.* **16** 624–6

Hadley G R 1992 Transparent boundary condition for the beam propagation method *IEEE J. Quantum Electron.* **28** 363–70

[37] Mişicu Ş, Rizea M and Greiner W 2001 Emission of electromagnetic radiation in α-decay *J. Phys. G: Nucl. Part. Phys.* **27** 993–1003

[38] Jackson D F and Rhoades-Brown M 1977 Theories of alpha-decay *Ann. Phys. N.Y.* **105** 151–86

[39] Fonda L, Ghirardi G C and Rimini A 1978 Decay theory of unstable quantum systems *Rep. Prog. Phys.* **41** 587–631

[40] Wilkinson S R *et al* 1997 Experimental evidence for non-exponential decay in quantum tunnelling *Nature* **387** 575–7

[41] Brändas E E and Elander N 1989 *Resonances* (Lecture Notes in Physics vol 325) (Berlin: Springer)

[42] RIPL-2 Reference Input Parameter Library *IEAE* http://www-nds.iaea.org/RIPL-2/

[43] Gavrila M 1992 Atomic structure and decay in high-frequency fields *Atoms in Intense Laser Fields* ed M Gavrila (Boston, MA: Academic) pp 435–510

[44] Mulser P and Bauer D 2010 *High Power Laser-Matter Interaction* (Springer Tracts in Modern Physics vol 238) (Heidelberg: Springer)

[45] Hofmann S 1996 Proton radioactivity *Nuclear Decay Modes* ed D N Poenaru (Bristol: IOP Publishing) ch 3 pp 143–203

[46] Åberg S, Semmes P B and Nazarewicz W 1997 Spherical proton emitters *Phys. Rev.* C **56** 1762

1998 *Phys. Rev.* C **58** 3011 Erratum

[47] Mişicu Ş and Cârjan N 1998 Proton decay from excited states in spherical nuclei *J. Phys. G: Nucl. Part. Phys.* **24** 1745–55

[48] Maglione E, Ferreira L S and Liotta R J 1999 Proton emission from deformed nuclei *Phys. Rev.* C **59** R589–92

[49] Davids C N and Esbensen H 2000 Decay rates of spherical and deformed proton emitters *Phys. Rev.* C **61** 054302

[50] Esbensen H and Davids C N 2000 Coupled-channels treatment of deformed proton emitters *Phys. Rev.* C **63** 014315

[51] Davids C N and Esbensen H 2004 Decay rate of triaxially deformed proton emitters *Phys. Rev.* C **69** 034314

[52] Eisenberg J M and Greiner W 1970 *Nuclear Theory, Volume II: Excitation Mechanisms of the Nucleus* (Amsterdam: North-Holland)

[53] Pashkevich V V 1971 On the asymmetric deformation of fissioning nuclei *Nucl. Phys.* A **169** 275–93

[54] Rizea M and Carjan N 2007 Bi-dimensional quantum tunneling by a time-dependent computational approach *Computer Physics Research Trends* ed S J Bianco (New York: NOVA Science) ch 5 pp 185–210

[55] Rizea M and Cârjan N 2008 Dynamical study of proton emission from deformed nuclei *Rom. Rep. Phys.* **60** 27–44

[56] Joachain C J, Kylstra N J and Potvliege R M 2012 *Atoms in Intense Laser Fields* (Cambridge: Cambridge University Press)

[57] Davids C N *et al* 1997 New proton radioactivities ^{165,166,167}Ir and ^{171}Au *Phys. Rev.* C **55** 2255–66
[58] Sonzogni A A 2002 Proton radioactivity in $Z > 50$ nuclides *Nucl. Data Sheets* **95** 1–48
[59] Flügge S 1971 *Practical Quantum Mechanics II* (Berlin: Springer)
[60] Davids C N *et al* 1998 Proton radioactivity from highly deformed nuclei *Phys. Rev. Lett.* **80** 1849–52
[61] Seweryniak V *et al* 2001 Rotational bands in the proton emitter ^{141}Ho *Phys. Rev. Lett.* **86** 1458–61
[62] Bingham C R 2005 Nuclear structure studies at the proton drip line via proton radioactivity studies *Nucl. Instrum. Methods* B **241** 185–9
[63] Catchelder J *et al* 1998 Observation of the exotic nucleus ^{145}Tm via its direct proton decay *Phys. Rev.* C **57** R1042–6
[64] Davids C N and Esbensen H 2001 Particle-vibration coupling in proton decay of near-spherical nuclei *Phys. Rev.* C **64** 034317
[65] Seweryniak D *et al* 2007 Effect of a triaxial nuclear shape on proton tunneling: the decay and structure of ^{145}Tm *Phys. Rev. Lett.* **99** 082502

Chapter 5

Heavy-ion scattering in a strong laser field

The study of laser-assisted collisions of atoms with electrons was investigated both theoretically (see [1, 2] and references therein) and experimentally [3]. The interest in carrying out such investigations was two-fold: to understand the relevance of plasma heating due to the simultaneous action of a laser pulse and an electron beam [4–8] and to access some of the electron–atom or electron–molecule scattering parameters that otherwise are not revealed [9–14].

Some of the theoretical aspects concerning the laser-assisted charge scattering deserve to be briefly mentioned. In an elastic scattering process between an electron and an atom, assisted by a laser field, the kinetic energy of the scattered electron E_f can be augmented/decreased with respect to the initial kinetic energy E_i, by multiples of photon energy $\hbar\omega$, i.e. $E_f - E_i = n\hbar\omega$. This process is called a 'free–free' transition and corresponds to *stimulated bremmsstrahlung* when $n < 0$, and to *inverse bremmsstrahlung* when $n > 0$. Such processes are current in astronomy in connection to the infrared opacity of the solar atmosphere and in plasma physics (*plasma heating*).

A sound theoretical framework aiming to describe an electron moving under the simultaneous action of an electromagnetic pulse and a scattering real potential was proposed independently by Kroll and Watson in [5] and Bunkin, Kazakov and Fedorov in [6]. In this context, Kroll and Watson derived the differential cross section of such a process in the case of a linearly polarized laser by taking into account the non-perturbative interaction between an electron and an atom accompanied by the exchange of n photons. The theorem established by these authors expresses $d\sigma_n/d\Omega$ as the simple product of the field-free elastic differential cross section, $d\sigma_{el}/d\Omega$, and the square of the Bessel function of order n, which encodes in its argument the field parameters. The importance of this theorem consists in providing a sound approximation to multiphoton energy transfers when the frequency of the laser wave is low or when the scattering potential is weak. It is worth noting that in a heavy-ion physics context only the first circumstance (low-frequency radiation) is valid since the scattering potentials are usually strong.

 © IOP Publishing Ltd 2023

Another feature, important for the present discussion, is that the Kroll–Watson theorem explains why quantum features recede in the laser-assisted scattering problem once multiphoton effects come into play and consequently classical aspects are highlighted. Experimental work carried out on laser-assisted electron–atom collision is nicely confirming the robustness of the Kroll–Watson approach [15, 16].

This chapter examines how the nucleus–nucleus potential scattering, in the presence of a strong electromagnetic field, unravels new classical and quantum features of the collision dynamics at the nuclear level. To the date, the influence of intense electromagnetic radiation on the scattering of protons and light nuclei on heavy-ion targets has been considered for the proton inelastic scattering on ^{12}C in monochromatic [17] and bichromatic fields [18], and the modification of elastic scattering differential cross sections in reactions of heavy ions with stable and loosely bound projectiles for arbitrary polarizations of the incident laser field [19, 20].

The theoretical framework required to analyze the heavy-ion scattering in an arbitrarily polarized radiation field and applications to the refractive scattering of loosely bound nuclei is discussed below.

5.1 Scattering in an arbitrarily polarized radiation field

Consider the simplest possible charged particle scattering colliding system, (Z_1, A_1) and (Z_2, A_2), with effective charge $Z_{\rm eff} = (Z_1A_2 - Z_2A_1)/(A_1 + A_2)$ in the field of a plane monochromatic electromagnetic wave of frequency ω and vector potential $\mathbf{A}$ [21]. The electromagnetic field is assumed to be spatially homogeneous and of arbitrary polarization specified by the unit vectors $\hat{\varepsilon}_1$, $\hat{\varepsilon}_2$ that span a plane orthogonal to the beam direction. The strength of the electric field component assumes the form [22]

$$\mathbf{E}(t) = \frac{1}{\sqrt{1+\xi^2}}\mathscr{E}_0(\hat{\varepsilon}_1 \sin\omega t + \xi\hat{\varepsilon}_2 \cos\omega t), \tag{5.1}$$

where $-1 \leqslant \xi \leqslant 1$ is the ellipticity parameter. For $\xi = 0$ one deals with a linearly polarized field, a case discussed in [19], whereas $\xi = \pm 1$ corresponds to the circular polarization case [20].

It was shown in chapter 1 that within the approximation of a space homogeneous field, the wave-function of the nucleus–nucleus relative motion satisfies the time-dependent Schrödinger equation (TDSE),

$$i\hbar\frac{\partial\psi(\mathbf{r}, t)}{\partial t} = [H_0 + H_{\rm int}(t)]\psi(\mathbf{r}, t), \tag{5.2}$$

where

$$H_0 = \frac{\mathbf{p}^2}{2\mu} + V(\mathbf{r}) \tag{5.3}$$

is the time-independent or field-free Hamiltonian of the two ions, of reduced mass $\mu = mA_1A_2/(A_1 + A_2)$, interacting via a two-body potential $V(\mathbf{r})$ in the absence of the laser field, whereas

$$H_{\text{int}}(t) = -\frac{e}{m}\left(\frac{Z_1}{A_1} - \frac{Z_2}{A_2}\right)\mathbf{p} \cdot \mathbf{A} + \frac{e^2}{2m}\left(\frac{Z_1^2}{A_1} + \frac{Z_2^2}{A_2}\right)\mathbf{A}^2 \tag{5.4}$$

is the time-dependent part of the total Hamiltonian, describing the interaction of the colliding system with the electromagnetic field. Performing a sequence of two unitary transformations on the wave-function,

$$\psi^{\text{K-H}}(\mathbf{r}, t) = \hat{U}_{\text{K-H}}\hat{U}_{\text{V}}\psi(\mathbf{r}, t), \tag{5.5}$$

one can simplify the TDSE equation, i.e. remove the terms $\sim\mathbf{A}^2$ and $\mathbf{p} \cdot \mathbf{A}$. The explicit expressions of $\hat{U}_{\text{V}}$ and $\hat{U}_{\text{K-H}}$ are given in chapter 1. In the Kramers–Henneberger representation, the TDSE assumes the form

$$\mathrm{i}\hbar\frac{\partial\psi^{\text{K-H}}(\mathbf{r}, t)}{\partial t} = \left[-\frac{\hbar^2}{2\mu}\nabla_r^2 + V(\mathbf{r} + \vec{\alpha})\right]\psi^{\text{K-H}}(\mathbf{r}, t), \tag{5.6}$$

where the classical displacement vector of the colliding system center-of-mass from its oscillation center of amplitude $\alpha_0 = -eZ_{\text{eff}}\mathcal{E}_0/\mu\omega^2$, when $\xi \neq 0$, is

$$\begin{aligned}\vec{\alpha} &= -\frac{eZ_{\text{eff}}}{\mu}\int_{-\infty}^{t} \mathrm{d}t'\mathbf{A}(t')\\ &= \frac{1}{\sqrt{1+\xi^2}}\alpha_0(\hat{\varepsilon}_1 \sin\omega t + \xi\hat{\varepsilon}_2 \cos\omega t).\end{aligned} \tag{5.7}$$

In the absence of the scattering potential the unperturbed wave-functions are

$$\psi_0^{\text{K-H}}(\mathbf{r}, t) = \Phi_{\mathbf{k}}(\mathbf{r})\mathrm{e}^{\frac{\mathrm{i}}{\hbar}E_k t}, \tag{5.8}$$

where

$$\Phi_{\mathbf{k}}(\mathbf{r}) = (2\pi)^{-3/2}\exp(\mathrm{i}\mathbf{k} \cdot \mathbf{r}) \tag{5.9}$$

are eigenstates of the non-perturbed Hamiltonian $H_0 = -\frac{\hbar^2}{2\mu}\nabla_r^2$. By switching to the interaction representation, the perturbed wave-function and the time-dependent scattering potential are transformed to

$$|\psi_I{}^{\text{K-H}}\rangle = \mathrm{e}^{\mathrm{i}H_0 t/\hbar}|\psi^{\text{K-H}}\rangle, \tag{5.10}$$

$$V_{\text{I}}(t) = \mathrm{e}^{\mathrm{i}H_0 t/\hbar}V(t)\mathrm{e}^{-\mathrm{i}H_0 t/\hbar}. \tag{5.11}$$

Thence the probability amplitude for the transition $\mathbf{k}_i \longrightarrow \mathbf{k}_f$ during the collision is presented in the form [23]

$$C_{\mathbf{k}_i,\mathbf{k}_f}(t) = \langle\Phi_{\mathbf{k}_f}|\mathcal{T}\exp\left[-\frac{\mathrm{i}}{\hbar}\int_0^{t} \mathrm{d}t \;\; V_I(t)(\mathbf{r} + \vec{\alpha})\right]|\Phi_{\mathbf{k}_i}\rangle, \tag{5.12}$$

where $\mathcal{T}$ is the Dyson time-ordering operator. For instance, when acting on the product of two operators which are functions of time, $\mathcal{T}$ moves the operator involving an earlier time to the right. If the limits of integration over time in equation (5.12) are extended to $t = -\infty$ for the initial time and $t = +\infty$ for the final time, the transition probability amplitude after the collision is denoted by the matrix element of the S-matrix,

$$C_{\mathbf{k}_i,\,\mathbf{k}_f}(\infty) = \langle \Phi_{\mathbf{k}_f} | \hat{S} | \Phi_{\mathbf{k}_i} \rangle. \tag{5.13}$$

A series expansion of the exponential in equation (5.12) in powers of $V_\mathrm{I}(t)$ provides the Born series. The first Born approximation reads

$$\begin{aligned} S_{fi}^{\mathrm{B}} &= S_{fi}^{(0)} + S_{fi}^{(1)} \\ &= \delta(\mathbf{k}_f - \mathbf{k}_i) + \frac{1}{\mathrm{i}\hbar} \langle \Phi_{\mathbf{k}_f} | \int_{-\infty}^{+\infty} \mathrm{d}t\, V_I(\mathbf{r} + \vec{\alpha}(t)) | \Phi_{\mathbf{k}_i} \rangle, \end{aligned} \tag{5.14}$$

where $\mathbf{k}_{i,\,f}$ are the vectors corresponding to the initial and final momenta of the colliding system in the CM (center of mass). Their absolute values are

$$k_i = \left(\frac{2\mu}{\hbar^2} E_{k_i}\right)^{1/2}, \quad k_f = \left(\frac{2\mu}{\hbar^2}(E_{k_i} + n\hbar\omega)\right)^{1/2}. \tag{5.15}$$

The above matrix element can be worked out by introducing the Fourier transform of the scattering potential in the K–H representation, i.e.

$$V(\mathbf{r} + \vec{\alpha}(t)) = \int \mathrm{d}\mathbf{q}\ \tilde{V}(\mathbf{q}) \exp[-\mathrm{i}\mathbf{q} \cdot (\mathbf{r} + \vec{\alpha}(t))]. \tag{5.16}$$

Using the integral representation of the delta function

$$\delta(\mathbf{k}) = \frac{1}{(2\pi)^3} \int \mathrm{d}\mathbf{r}\ \exp(\mathrm{i}\mathbf{k} \cdot \mathbf{r}), \tag{5.17}$$

it is then easy to infer that

$$\langle \Phi_{\mathbf{k}_f} | V(\mathbf{r} + \vec{\alpha}(t)) | \Phi_{\mathbf{k}_i} \rangle = \tilde{V}(\mathbf{k}_i - \mathbf{k}_f) \mathrm{e}^{-\mathrm{i}(\mathbf{k}_i - \mathbf{k}_f)\cdot\vec{\alpha}(t)}. \tag{5.18}$$

As pointed out in the appendix, the generating function of the Bessel functions of integer order reads

$$\mathrm{e}^{-\mathrm{i}\Delta\mathbf{q}\cdot\vec{\alpha}(t)} = \sum_{n=-\infty}^{\infty} J_n(\mathscr{Z}) \mathrm{e}^{-\mathrm{i}n(\omega t + \Psi)}, \tag{5.19}$$

where

$$\mathscr{Z} = \frac{\alpha_0}{\sqrt{1 + \xi^2}} \sqrt{(\Delta\mathbf{q} \cdot \hat{\varepsilon}_1)^2 + (\xi \Delta\mathbf{q} \cdot \hat{\varepsilon}_2)^2}. \tag{5.20}$$

If one denotes by $\chi_{1,2}$ the angles between the momentum transfer $\Delta\mathbf{q}$ and the polarization unit vectors $\hat{\varepsilon}_{1,2}$, the phase Ψ in equation (5.19) can be expressed as

$$\Psi = \mathrm{atan}\left(\frac{\cos\chi_1}{\cos\chi_2}\right). \tag{5.21}$$

Then, the second term in the rhs of equation (5.14) can be put in the form

$$\langle\Phi_{\mathbf{k}_f}|\int_{-\infty}^{+\infty}\mathrm{d}t\ \ V_I(\mathbf{r}+\vec{\alpha}(t))|\Phi_{\mathbf{k}_i}\rangle = \tilde{V}(\mathbf{q})\sum_{n=-\infty}^{\infty}J_n(\mathscr{Z})\int_{-\infty}^{+\infty}\mathrm{d}t e^{\frac{\mathrm{i}}{\hbar}(E_{k_f}-E_{k_i}-n\hbar\omega)t}. \tag{5.22}$$

Since

$$\delta(E_{k_f}-E_{k_i}-n\hbar\omega) = \frac{1}{2\pi}\int_{-\infty}^{+\infty}\mathrm{d}t\exp\left\{\frac{\mathrm{i}}{\hbar}(E_{k_f}-E_{k_i}-n\hbar\omega)t\right\}, \tag{5.23}$$

the first-order matrix element of S is finally equated to

$$\begin{aligned} S_{fi}^{(1)} &\equiv \langle\Phi_{\mathbf{k}_f}|\int_{-\infty}^{+\infty}\mathrm{d}t\ \ V_I(\mathbf{r}+\vec{\alpha}(t))|\Phi_{\mathbf{k}_i}\rangle \\ &= -2\pi\mathrm{i}\tilde{V}(\mathbf{q})\sum_{n=-\infty}^{\infty}J_n(\mathscr{Z})\delta(E_{k_f}-E_{k_i}-n\hbar\omega). \end{aligned} \tag{5.24}$$

The energy balance is encoded in the delta function, $\delta(E_{k_f}-E_{k_i}-n\hbar\omega)$, entering the sum in the rhs of equation (5.24),

$$E_{k_f} = E_{k_i} + n\hbar\omega, \tag{5.25}$$

with $n = 0, \pm1, \pm2, \ldots$. Positive values ($n > 0$) signify absorption, dubbed as *inverse* bremsstrahlung, thus $E_{k_f} > E_{k_i}$. Negative values ($n < 0$) signify emission or *stimulated* bremsstrahlung. In this last case the number of emitted photons in the presence of the laser field has a limiting value, i.e.

$$-n \leqslant \frac{E_{k_i}}{\hbar\omega}. \tag{5.26}$$

Within the conventional scattering theory [24] the S-matrix is related to the matrix element of the transition operator $\hat{T}$,

$$\hat{S} = \hat{I} - 2\pi\mathrm{i}\hat{T}, \tag{5.27}$$

where $\hat{I}$ is the unit matrix. Introducing the Born approximation for the T-matrix of the conventional scattering process accompanied by the exchange of n photons,

$$T_{fi}^{\mathrm{B}}(n) = J_n(\mathscr{Z})\mathrm{e}^{-\mathrm{i}n\Psi}\tilde{V}(\Delta\mathbf{q}), \tag{5.28}$$

and substituting (5.19) and (5.16) in (5.24), the first-order contribution to the Born S-matrix element reads

$$S_{fi}^{(1)} = -2\pi\mathrm{i}\sum_{n=-\infty}^{\infty}\delta(E_{k_f}-E_{k_i}-n\hbar\omega)T_{fi}^{\mathrm{B}}(n). \tag{5.29}$$

Then, in a manner completely analogous to the field-free case [24], the differential cross section for the nucleus–nucleus scattering, accompanied by the absorption (emission) of n photons, is given by

$$\frac{\mathrm{d}\sigma_n}{\mathrm{d}\Omega} = \left(\frac{\mu}{2\pi\hbar^2}\right)^2 \frac{k_f}{k_i} |T^{\mathrm{B}}_{fi}(n)|^2. \tag{5.30}$$

Recalling the expression of the field-free elastic scattering differential cross section in the first Born approximation [26]

$$\left(\frac{\mathrm{d}\sigma}{\mathrm{d}\Omega}\right)_{\mathrm{B1}}(\Delta\mathbf{q}) = \left(\frac{\mu}{2\pi\hbar^2}\right)^2 |\tilde{V}(\mathbf{q})|^2, \tag{5.31}$$

the ratio of the n-photon to the Born elastic differential cross section reads

$$\left(\frac{\mathrm{d}\sigma_n}{\mathrm{d}\sigma_{\mathrm{B1}}}\right)(\Delta\mathbf{q}) = \frac{k_f}{k_i} J_n^2\left(\frac{\alpha_0 \Delta q}{\sqrt{1+\xi^2}} \sqrt{(\cos\chi_1)^2 + \xi^2(\cos\chi_2)^2}\right). \tag{5.32}$$

As can be seen in equation (5.32) the differential cross section depends, via the argument of the Bessel function, on the orientation of the polarization vectors $\hat{\varepsilon}_{1,2}$ relative to $\Delta\mathbf{q}$. The effect of the laser fields enters the expression (5.32) in a non-perturbative way, via the Bessel function $J_n^2(\mathscr{Z})$.

It is worthwhile to comment on the case when there is a huge discrepancy in magnitude between the collision energy and the energy of exchanged photons, i.e. $E_{k_i} \gg n\hbar\omega$. From equation (5.15) it results that $k_f \approx k_i$. Consequently, summing equation (5.32) after the photon number, such that always $k_f \geqslant 0$, and using the addition theorem for Bessel function of integer order (9.1.76 of [25]),

$$1 = \sum_n J_n^2(\mathscr{Z}), \tag{5.33}$$

the following sum-rule is fulfilled

$$\sum_n \frac{\mathrm{d}\sigma_n}{\mathrm{d}\Omega} \approx \left(\frac{\mathrm{d}\sigma}{\mathrm{d}\Omega}\right)_{\mathrm{B1}}. \tag{5.34}$$

The sum-rule, derived in the first Born approximation, was claimed by Kroll and Watson to be valid to all orders within the low-frequency approximation (LFA) [5]. This point was disputed following the numerical experiments reported in [27].

The applicability of the Born approximation is insured provided that condition

$$\sqrt{\frac{\mu}{2E_{k_i}}} \frac{Z_1 Z_2 e^2}{\hbar} \ll 1 \tag{5.35}$$

is fulfilled. As for the range of applicability of equation (5.32), it is worth mentioning that although this issue was never rigorously investigated (see [6]), a condition leading to equation (5.32) is that the laser frequency is small compared to the bombarding energy, i.e.

$$\xi = \frac{\hbar\omega}{2E_{k_i}} \ll 1. \tag{5.36}$$

The discussion continues by considering two geometries. In both configurations the unit polarization vector $\hat{\varepsilon}_1$ is oriented along the z-axis whereas $\hat{\varepsilon}_2$ points in the positive direction of the y-axis. The initial momentum of the heavy-ion beam and $\hat{\varepsilon}_1$ are collinear ($\mathbf{k}_i \| \hat{\varepsilon}_1$) in the first geometry and perpendicular ($\mathbf{k}_i \perp \hat{\varepsilon}_1$) in the second one. The scattered beam, specified by the final momentum $\mathbf{k}_f$, spans an angle Θ from the z-axis, whereas its projection on the x–y plane forms and angle Φ with the x-axis. It is easy to observe that the angles $\chi_{1,2}$ are expressed in terms of the polar angles ($\Theta \equiv \theta_{\text{c.m.}}, \Phi$) and having in mind that $n\hbar\omega \ll E_{k_i}$ ($k_i \approx k_f$), for all relevant values of n, it follows from equation (5.15) that

$$\frac{\cos\chi_1}{\cos\chi_2} \approx \frac{\cot\theta_{\text{c.m.}} - \csc\theta_{\text{c.m.}}}{\sin\Phi} \ (\mathbf{k}_i \| \hat{\varepsilon}_1) \ ; \ \frac{\cos\chi_1}{\cos\chi_2} \approx \frac{\cot\theta_{\text{c.m.}}}{\sin\Phi} \ (\mathbf{k}_i \perp \hat{\varepsilon}_1). \tag{5.37}$$

Hence, for linear polarization ($\xi = 0$) the n-photon to field-free Born differential cross section is

$$\left(\frac{\mathrm{d}\sigma_n}{\mathrm{d}\sigma_{\text{B1}}}\right)(k_i, \theta_{\text{c.m.}}, \Phi) \approx J_n^2(2\alpha_0 k_i \sin^2\theta_{\text{c.m.}}/2) \quad (\mathbf{k}_i \| \hat{\varepsilon}_1) \tag{5.38}$$

$$\approx J_n^2(\alpha_0 k_i \cos\theta_{\text{c.m.}}) \quad (\mathbf{k}_i \perp \hat{\varepsilon}_1). \tag{5.39}$$

For circular polarization ($\xi = \pm 1$)

$$\left(\frac{\mathrm{d}\sigma_n}{d\sigma_{\text{B1}}}\right)(k_i, \theta_{\text{c.m.}}, \Phi) \approx J_n^2\left(\sqrt{2}\,\alpha_0 k_i \sin^2\frac{\theta_{\text{c.m.}}}{2}\sqrt{1 + \cot^2\frac{\theta_{\text{c.m.}}}{2}\sin^2\Phi}\right)(\mathbf{k}_i \| \hat{\varepsilon}_1) \tag{5.40}$$

$$\approx J_n^2\left(\frac{\alpha_0 k_i}{\sqrt{2}}\sqrt{\cos^2\theta_{\text{c.m.}}\cos^2\Phi + \sin^2\Phi}\right) \ (\mathbf{k}_i \perp \hat{\varepsilon}_1) \tag{5.41}$$

5.2 Refractive scattering of loosely bound nuclei

The framework developed above is next applied to the case of a Ti:sapphire laser beam ($\hbar\omega = 1.5$ eV) impinging on a projectile–target system.

In figures 5.1–5.4 the laser-assisted cross section as a ratio to the Rutherford cross section ($\mathrm{d}\sigma_n/\mathrm{d}\sigma_{\text{R}}$) is displayed; also displayed are its F/N decompositions (total, far- and nearside) in the photonless case ($n = 0$) and for the absorption of $n = 1$ and $n = 2$ photons for two moderate intensities ($I = 10^{11}$ and 10^{13} W cm^{-2}) and two orientations of the electromagnetic field with respect to the initial ion beam. The case of a linearly polarized laser field is represented in figures 5.1 and 5.3, and that for circular polarization ($\xi = 1$) in figures 5.2 and 5.4. For a linearly polarized radiation of intensity $I = 10^{11}$ W cm^{-2} the shallow Airy structure, present in the field-free case for the reaction ^{6}Li + ^{7}Li, is changed in the case of orientation $\mathbf{k}_i \| \hat{\varepsilon}_1$ by a deeper structure, shifted in the forward direction for $n = 0$, that subsequently vanishes for the higher

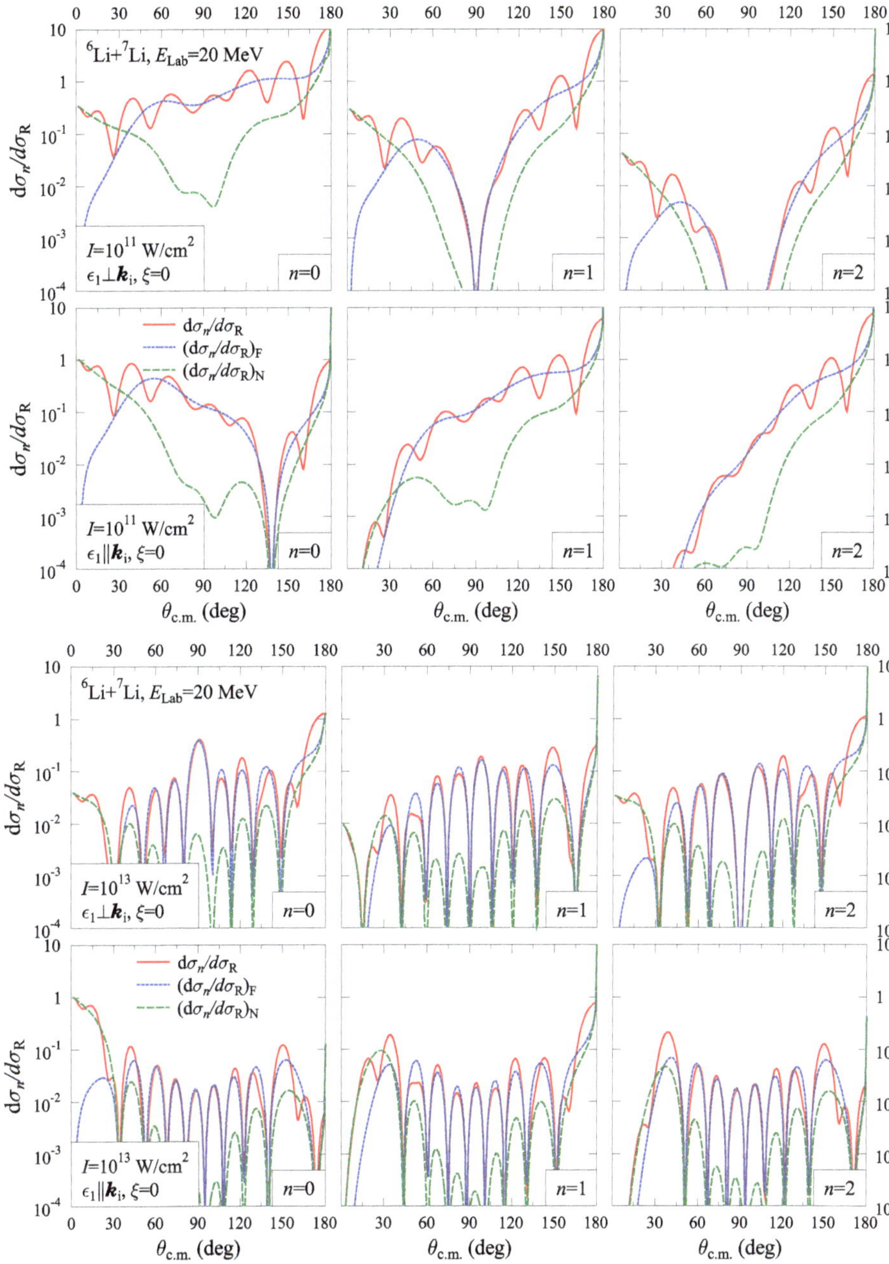

Figure 5.1. Elastic scattering ^{6}Li+^{7}Li at $E = 20$ MeV assisted by a linearly polarized ($\xi = 0$) laser field with photon energy $\hbar\omega = 1.5$ eV and for two intensities: $I = 10^{11}$ W cm^{-2} (top panels) and $I = 10^{13}$ W cm^{-2} (bottom panels). Two geometric configurations (orientation of the electromagnetic field with respect to the initial ion beam) are selected: $\mathbf{k}_i \| \hat{\varepsilon}_1$ and $\mathbf{k}_i \perp \hat{\varepsilon}_1$. $d\sigma_n/d\sigma_{B1}$ is calculated for $n = 0, 1, 2$ absorbed photons. The F/N decomposition is figured by the short-dashed curve (farside) and the long-dashed curve (nearside). Reproduced with permission from [20]. Copyright 2022 the American Physical Society.

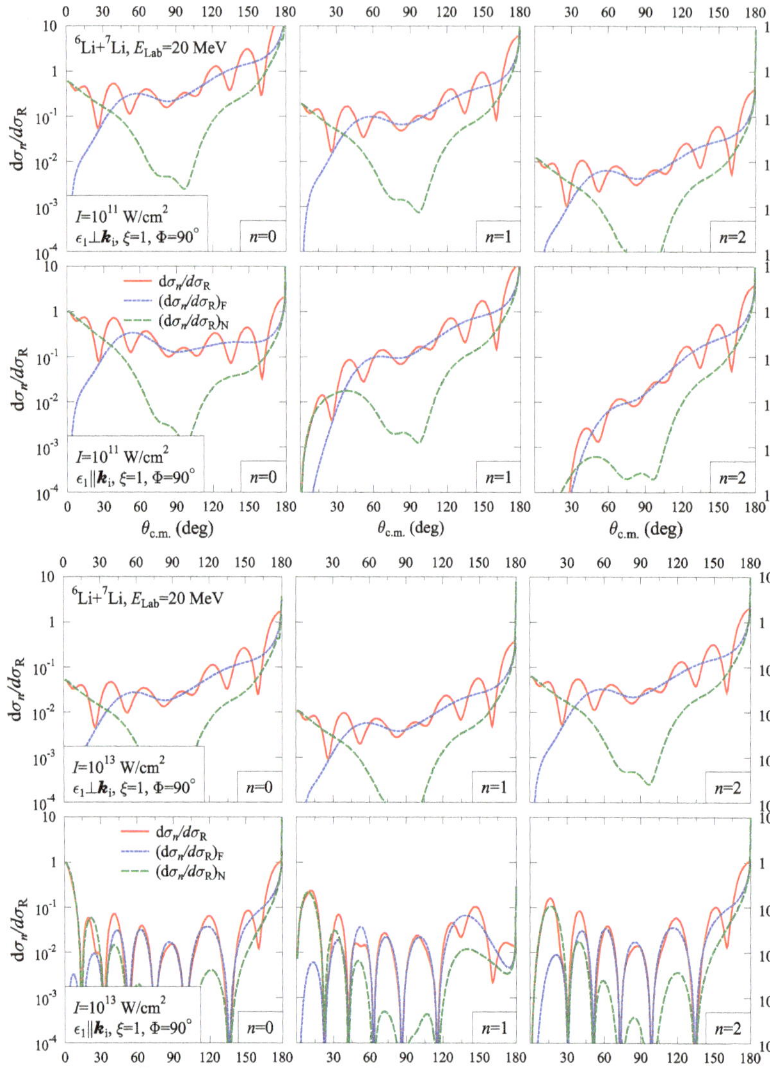

Figure 5.2. Elastic scattering ^{6}Li + ^{7}Li at $E = 20$ MeV assisted by a circularly polarized ($\xi = 1$) laser field with photon energy $\hbar\omega = 1.5$ eV, at two intensities: $I = 10^{11}$ W cm^{-2} (top panels) and $I = 10^{13}$ W cm^{-2} (bottom panels). Two geometric configurations (orientation of the electromagnetic field with respect to the initial ion beam) are selected: $\mathbf{k}_i \| \hat{\varepsilon}_1$ (bottom panels) and $\mathbf{k}_i \perp \hat{\varepsilon}_1$ (top panels) and azimuth's orientation $\Phi = 90°$. The F/N decomposition is figured by the short-dashed curve (farside) and the long-dashed curve (nearside). Reproduced with permission from [20]. Copyright 2022 the American Physical Society.

intensity. Instead, at $I = 10^{13}$ W cm^{-2}, pronounced minima appear for $n = 2, 3$ for the orientation $\mathbf{k}_i \perp \hat{\varepsilon}_1$. The pronounced Airy structure, still visible at the lowest considered intensity, is significantly altered at higher intensity when large amplitude oscillations are superimposed on it with Fraunhofer characteristic oscillations still visible in the forward angle sector. However, this behavior is not observed in any of the two

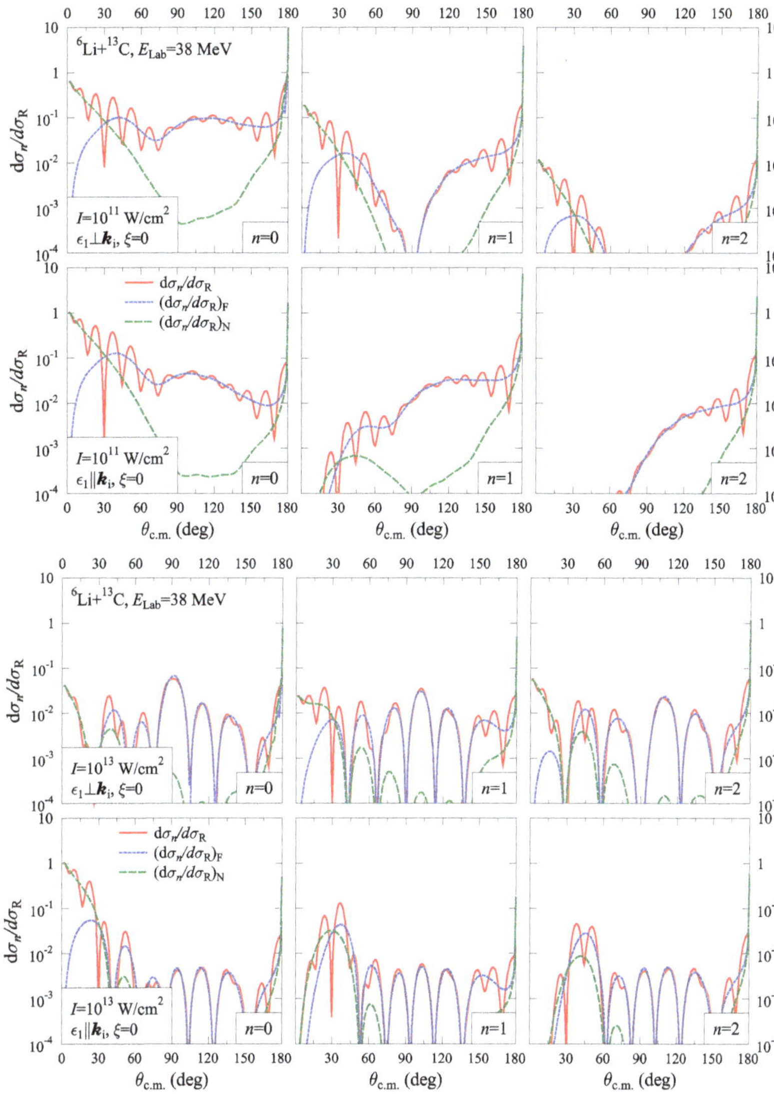

Figure 5.3. The same as in figure 5.1 for the elastic scattering ^{6}Li + ^{13}C at E = 38 MeV. Reproduced with permission from [20]. Copyright 2022 the American Physical Society.

reactions, for circular polarization and orientation $\mathbf{k}_i \perp \hat{\epsilon}_1$. Here the refractive and diffractive structures are preserved at $I = 10^{13}$ W cm^{-2}.

By increasing I and therefore stimulating the exchange of photons between the projectile–target system and the electromagnetic fields determines overall a depletion of the cross sections. In the same time $d\sigma_{n>0}/d\Omega$ overruns in most cases $d\sigma_0/d\Omega$.

The role played by the ellipticity can be inferred from the above mentioned figures. When circular polarized light is considered, an enhancement of $\sigma_{n=1,2}$, but not also of $\sigma_{n=0}$, takes place for the orientation $\mathbf{k}_i \perp \hat{\epsilon}_1$.

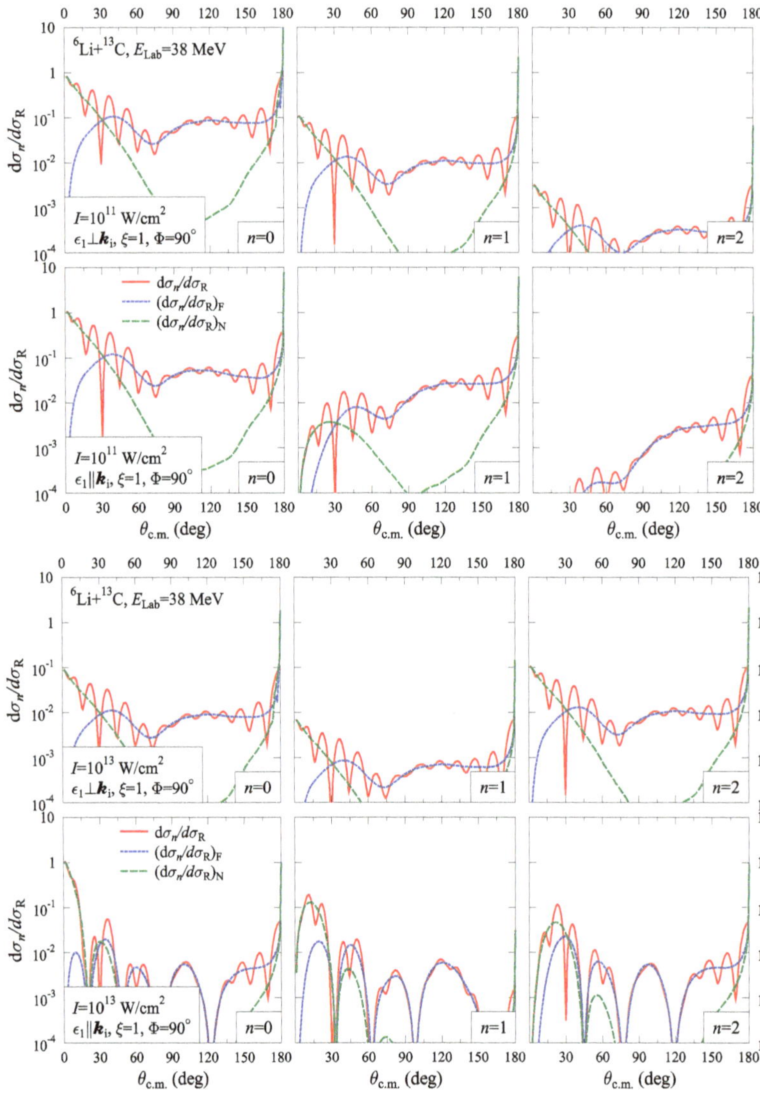

Figure 5.4. The same as in figure 5.2 for the elastic scattering ^{6}Li + ^{13}C at E = 38 MeV. Reproduced with permission from [20]. Copyright 2022 the American Physical Society.

The effect of multiphoton absorption at various scattering angles $\theta_{\rm c.m.}$ can be estimated by representing the n-photon strengths $d\sigma_n/d\sigma_{el}$, normalized according to (5.34) in terms of the number of exchanged photons (see [19]) and is displayed in figures 5.5–5.8. At low scattering angle ($\theta_{\rm c.m.} = 5°$), the multiphoton absorption effects are displayed in figure 5.5 for ^{6}Li + ^{7}Li (E = 20 MeV) and in figure 5.7 for ^{6}Li + ^{13}C (E = 38 MeV) for two laser intensities, for linear and circular polarizations and the scattering geometries chosen as before. For the $\mathbf{k}_i \perp \hat{\epsilon}_1$ orientation, the photonless component dominates the strength distribution at the lower intensity.

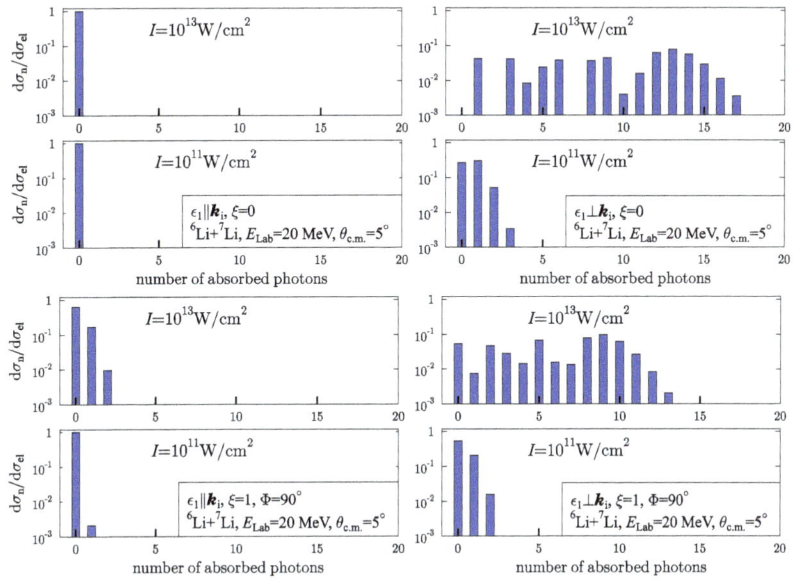

Figure 5.5. The n-photon differential cross section for the elastic scattering reaction ^{6}Li + ^{7}Li at $E_{\mathrm{lab}} = 20$ MeV as a function of the number of absorbed photons for two laser intensities ($I = 10^{11}$ and 10^{13} W cm^{-2}), linear polarization ($\xi = 0$, the four panels from above), and circular polarization ($\xi = 1$, the four panels from below), scattering angle $\theta_{\mathrm{c.m.}} = 5°$ and for two geometric configurations: $\mathbf{k}_i \| \hat{\varepsilon}_1$ (left panels) and $\mathbf{k}_i \perp \hat{\varepsilon}_1$ (right panels). Reproduced with permission from [20]. Copyright 2022 the American Physical Society.

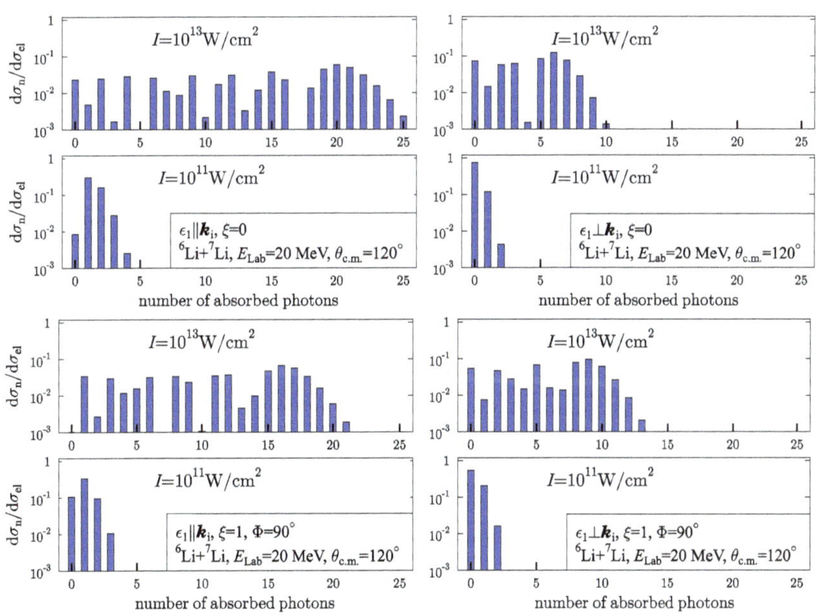

Figure 5.6. The same as in figure 5.5 for the scattering angle $\theta_{\mathrm{c.m.}} = 120°$. Reproduced with permission from [20]. Copyright 2022 the American Physical Society.

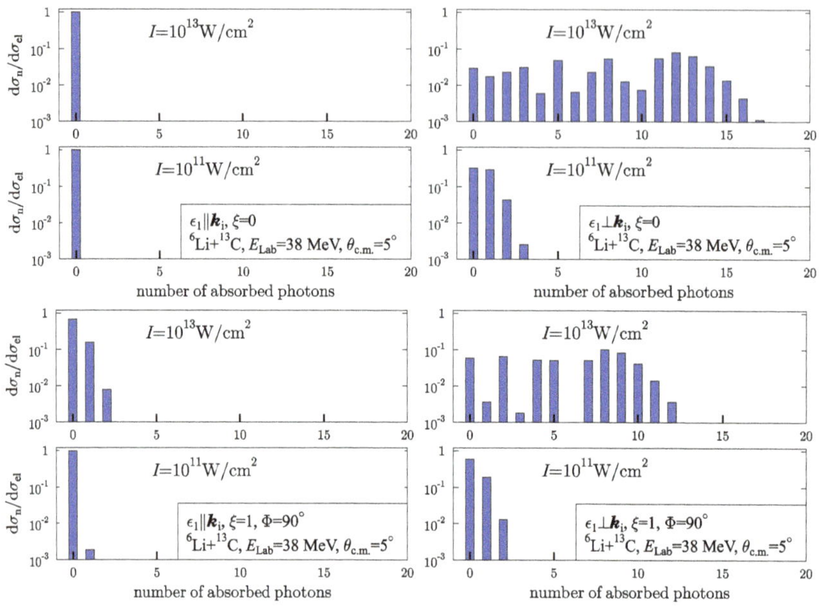

Figure 5.7. The same as in figure 5.5 for the reaction ^{6}Li + ^{13}C at $E_{\text{lab}} = 38$ MeV. Reproduced with permission from [20]. Copyright 2022 the American Physical Society.

For higher intensity strengths, components with $n \geqslant 1$ take over in importance. Instead, for the $\mathbf{k}_i \| \hat{\varepsilon}_1$ orientation the multiphoton effect is very weak at forward angles.

The case of a large scattering angle ($\theta_{\text{c.m.}} = 120°$) is displayed in figures 5.6 and 5.8. In this case multiphoton processes are already initiated at the lowest intensity for the orientation $\mathbf{k}_i \| \hat{\varepsilon}_1$.

The analysis presented above is confined to the inverse *bremsstrahlung* case, i.e. photon absorption. By symmetry ($J_n^2 = J_{-n}^2$), the case of photon emission leads to identical results.

5.3 Elastic scattering in a linearly polarized short laser pulse

For shorter laser pulses such as the ultra-short pulses lasting few optical cycles, the electromagnetic wave can be formed by superimposing monochromatic plane waves. Specializing to non-dispersive, linearly polarized laser pulses, the vector potential reads [28]

$$\mathbf{A}(\mathbf{r}, t) = \hat{\varepsilon} A_0 \int_{-\infty}^{t} F(\omega t' - \mathbf{k} \cdot \mathbf{r}) \sin(\omega t' - \mathbf{k} \cdot \mathbf{r}) \mathrm{d}t', \tag{5.42}$$

where $\hat{\varepsilon}$ is the polarization vector, A_0 is related to the amplitude of the electric field, i.e. $\mathscr{E}_0 = -\omega A_0$, and F is a nonnegative function that describes the shape (envelope) of the pulse. In the dipole approximation the vector potential satisfies the Coulomb gauge, $\nabla \cdot \mathbf{A} = 0$, and with no sources present, the electric field of the laser pulse is

$$\mathbf{E} = -\frac{\partial \mathbf{A}}{\partial t} = \hat{\varepsilon} \mathscr{E}_0 F(\omega t - \mathbf{k} \cdot \mathbf{r}) \sin(\omega t - \mathbf{k} \cdot \mathbf{r}). \tag{5.43}$$

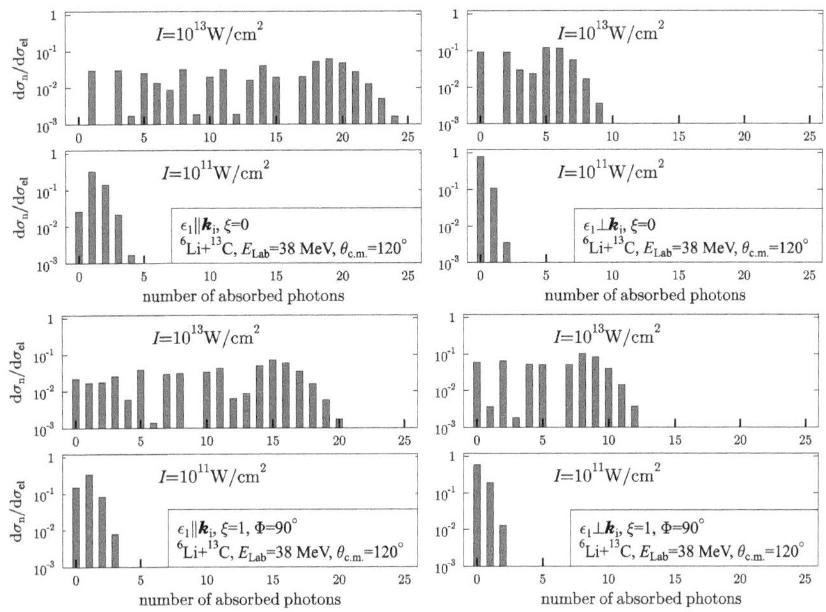

Figure 5.8. The same as in figure 5.6 for the reaction ^{6}Li + ^{13}C at $E_{\text{lab}} = 38$ MeV. Reproduced with permission from [20]. Copyright 2022 the American Physical Society.

In the dipole approximation the electric field of a linearly polarized laser pulse is

$$\mathbf{E}(t) = \hat{\varepsilon}\mathscr{E}_0 F(t) \sin \omega t. \tag{5.44}$$

Introducing the pulse duration τ_p as the full width at half maximum of the intensity profile $|\mathbf{E}(t)|^2$, three types of envelopes that are commonly encountered in the literature (see [29]). They were already defined in chapter 1: rectangular (step function)

$$F(t) = \theta(\tau_p - t), \tag{5.45}$$

sine-squared

$$F(t) = \sin^2\!\left(\frac{\pi t}{\tau_p}\right)\theta(\tau_p - t), \tag{5.46}$$

and Gaussian

$$F(t) = \exp\left[-4\pi^2 t^2/\tau_p^2\right]\theta(\tau_p - t). \tag{5.47}$$

The pulse duration is taken as an integer multiple of the laser period, i.e. $\tau_p = NT$. Although for these pulses the minimum N is about a few dozens of cycles, it is instructive to consider also pulses a few cycles in length.

In order to adapt the K–W theorem (5.32) to the case of laser pulses of finite duration we make use of the suggestion by Krüger and Jung [30]. It is assumed that the shape function $F(t)$ varies so slowly with t, that for time intervals long compared to $1/\omega$ it can be approximated piecewise to a good accuracy by the sum

$$F(t) = \sum_{p=1}^{N} f_p[\theta(T_p - t) - \theta(T_{p-1} - t)], \tag{5.48}$$

where

$$f_p = F\left(\frac{T_{p-1} + T_p}{2}\right). \tag{5.49}$$

Consequently, the differential cross section for n photons exchanged is obtained as a time average over the interval $[T_0, T_N]$ of length $\tau_p = T_N - T_0$,

$$\frac{d\bar{\sigma}_n}{d\Omega} = \frac{1}{\tau_p}\int_{T_0}^{T_N} dt \frac{d\sigma_n(F(t))}{d\Omega}, \tag{5.50}$$

where $(d\sigma_n(F(t))/d\Omega)$ is calculated for $F(t) = f_p$. Eventually the following compact expression is obtained:

$$\frac{d\bar{\sigma}_n}{d\Omega} \approx \frac{k_f}{k_i}\left(\frac{d\sigma}{d\Omega}\right)_{\text{elast}} \frac{1}{\tau_p}\int_{T_0}^{T_n} dt J_n^2(\mathbf{k} \cdot \vec{\alpha}_0(t)), \tag{5.51}$$

where

$$\alpha_0(t) = -\frac{eZ_{\text{eff}} f_p \mathscr{E}_0}{\mu\omega^2}[\theta(t - T_{p-1}) - \theta(T_p - t)]. \tag{5.52}$$

In the left panel of figure 5.9 the strengths (5.51) are displayed for the three modulated signals discussed above with pulse duration $\tau_p = 2\pi/\omega$ and intensity $I = 10^{17}$ W cm^{-2}, at small angle ($\theta_{\text{c.m.}} = 5°$). The dependence of the strengths on the

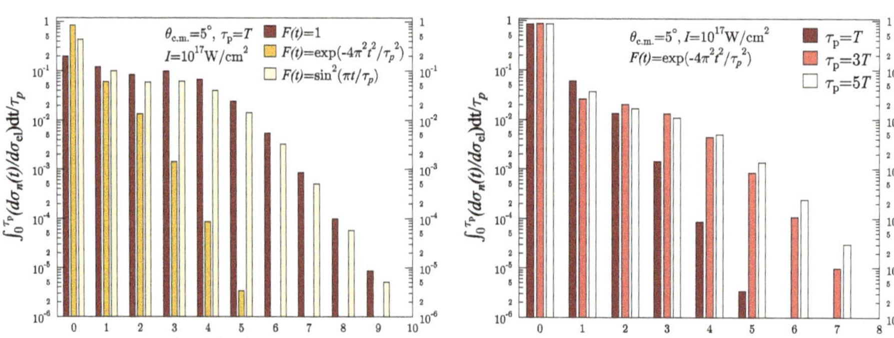

Figure 5.9. Left panel: The time average of the n-photon differential cross section (5.50) over the pulse duration τ_p as a function of the number of absorbed photons for three envelopes of the electric field $F(t)$ (square, Gaussian, and sine-squared), intensity $I = 10^{17}$ W cm^{-2}, duration, $\tau_p = T$ and fixed scattering angle ($\theta_{\text{c.m.}} = 5°$) for the Ti:sapphire laser. Right panel: The time average of the n-photon differential cross section (5.50) over the pulse duration τ_p as a function of the number of absorbed photons, for a Gaussian envelope and intensity $I = 10^{17}$ W cm^{-2} at fixed scattering angle ($\theta_{\text{c.m.}} = 5°$) for the Ti:sapphire laser. Three pulse durations are selected: $\tau_p = T$, $3T$, and $5T$. Reproduced with permission from [19]. Copyright 2022 Springer Nature.

shape of the chosen modulated pulse is striking. For the Gaussian pulse, the strength drops fast with n and consequently the 'multiphoton' contributions are manifest for the square pulse up to $n = 4$. The strengths for the sine-squared pulse display a behavior closer to the rectangular pulse.

In the right panel of figure 5.9 the same quantity is plotted, this time only for the Gaussian pulse and for finite pulse durations ($\tau_p = T$, $3T$, and $5T$). An expected outcome is that by increasing τ_p, the exchange of photons is enhanced.

References

[1] Mittleman M 1993 *Introduction to the Theory of Laser-Atom Interactions* 2nd edn (New York: Plenum)

[2] Milošević D B and Ehlotzky F 2003 Scattering and reaction processes in powerful laser fields *Adv. At. Mol. Opt. Phys.* **49** 373–532

[3] Kanya R, Morimoto Y and Yamanouchi K 2014 Laser-assisted electron scattering and diffraction in ultrashort intense laser fields *Progress in Ultrafast Intense Laser Science X* (Springer in Chemical Physics vol 106) ed K Yamanouchi *et al* (Cham: Springer International) pp 1–16

[4] Marcuse D 1962 Stimulated emission of bremsstrahlung *Bell. Syst. Tech. J.* **41** 1557–71

[5] Kroll N M and Watson K M 1973 Charge particle scattering in the presence of a strong electromagnetic wave *Phys. Rev.* A **8** 804–9

[6] Bunkin F V, Kazakov A E and Fedorov M V 1972 Interaction of intense optical radiation with free electrons *Usp. Fiz. Nauk* **107** 559–93
1973 *Sov. Phys. Usp.* **15** 416–35

[7] Seely J F and Harris E G 1973 Heating of a plasma by multiphoton inverse bremsstrahlung *Phys. Rev.* A **7** 1064–7

[8] Henneberger W C and Shatas R A 1977 Effect of electron–electron collisions on stimulated multiphoton *Phys. Rev.* A **15** 2501–4

[9] Mason N J 1989 Laser-assisted electron–atom collisions *Contemp. Phys.* **30** 449–63

[10] Wallbank B and Holmes J K 1993 Laser-assisted electron–atom collisions *Phys. Rev.* A **48** R2515–8

[11] Wallbank B and Holmes J K 2001 Laser-assisted elastic electron scattering from helium *Can. J. Phys.* **79** 1237–46

[12] Ehlotzky F 1978 X-ray scattering in the presence of a strong laser field *Phys. Lett.* B **25** 221–5

[13] Ehlotzky F 1981 Scattering phenomena in strong radiation fields I *Can. J. Phys.* **59** 1200–11

[14] Ehlotzky F 1984 Scattering phenomena in strong radiation fields II *Can. J. Phys.* **63** 907–32

[15] Musa M O, MacDonald A, Tidswell L, Holmes J and Wallbank B 2010 Laser induced free–free transitions in elastic electron scattering from CO_2 *J. Phys. B: At. Mol. Opt. Phys.* **43** 175201

[16] de Harak B A *et al* 2016 Effects of polarization direction on laser-assisted free-free scattering *Plasma Sources Sci. Technol.* **25** 035201

[17] Barna I F and Varró S 2015 Laser-assisted proton collision on light nuclei at moderate energies *Laser Part. Beams* **33** 299–305

[18] Barna I F and Varró S 2016 Proton scattering on carbon nuclei in bichromatic laser field at moderate energies *Nucl. Instrum. Methods Phys.* B **369** 77–82

[19] Mişicu Ş and Carstoiu F 2018 Fraunhofer and refractive scattering of heavy ions in strong laser fields *Eur. Phys. J.* A **54** 90

[20] Misicu Ş 2022 Refractive scattering of loosely bound nuclei in arbitrarily polarized laser fields *Phys. Rev.* C **106** 034612

[21] Mişicu Ş and Rizea M 2013 α-decay in ultra-intense laser fields *J. Phys. G: Nucl. Part. Phys.* **40** 095101

[22] Gavrila M 1992 Atomic structure and decay in high-frequency fields *Atoms in Intense Laser Fields* ed M Gavrila (Boston, MA: Academic)

[23] Alder K and Winther A 1975 *Electromagnetic Excitation: Theory of Coulomb Excitation with Heavy Ions* (Amsterdam: North-Holland)

[24] Messiah A 1960 *Mécanique Quantique* (Paris: Tome II)

[25] Abramovitz M and Stegun I 1962 *Handbook of Mathematical Functions* (Washington, DC: National Bureau of Standards)

[26] Joachain C J 1975 *Quantum Collision Theory* (Amsterdam: North-Holland)

[27] Daniele R, Ferrante G, Morales F and Trombetta F 1986 On the validity of the sum rule for multiphoton free-free transitions *J. Phys. B: At. Mol. Phys.* **19** L133–9

[28] Joachain C J, Kylstra N J and Potvliege R M 2012 *Atoms in Intense Laser Fields* (Cambridge: Cambridge University Press)

[29] Diels J-C and Rudolph W 2006 *Ultrashort Laser Pulse Phenomena* 2nd edn (Amsterdam: Elsevier)

[30] Krüger H and Jung C 1978 Low-frequency approach to multiphoton free–free transitions induced by realistic laser pulses *Phys. Rev.* A **17** 1706–12

IOP Publishing

Nuclear Dynamics in Strong Laser Fields

Şerban Mişicu

Appendix A

A.1 Classical motion in a rapidly oscillating field [1]

Consider an external force which varies periodically in time at a high frequency ($\omega \gg \tau^{-1}$) and acts upon a particle of mass m described by the trajectory $x(t)$ and which has associated a time scale τ when it moves in the unperturbed potential U along the trajectory $X(t)$, i.e.

$$f(t) = f_1 \cos \omega t + f_2 \sin \omega t. \tag{A.1}$$

The Lagrangian of the particle in the external field reads

$$\mathscr{L} = \frac{1}{2} m\dot{x}^2(t) - V(x) + xf(t). \tag{A.2}$$

The Euler–Lagrange equations provide the perturbed equation of motion

$$m\ddot{x}(t) = -\frac{\mathrm{d}V}{\mathrm{d}x} + f(t). \tag{A.3}$$

Denoting by $X(t)$ the non-perturbed trajectory, the equation motion reads

$$m\ddot{X} = -\frac{\mathrm{d}V}{\mathrm{d}X}. \tag{A.4}$$

Assuming next that the motion consists of small rapid oscillations, $\alpha(t)$, of frequency ω superimposed on the unperturbed trajectory

$$x(t) \approx X(t) + \alpha(t), \tag{A.5}$$

the time-average of the dynamical variables over an interval T is

$$\langle X(t) \rangle_T = \frac{1}{T} \int_{t-T/2}^{t+T/2} X(t) \mathrm{d}t. \tag{A.6}$$

It is also assumed that $X(t)$ varies a little during a period, i.e.

$$\langle X(t) \rangle_T \approx X(t). \tag{A.7}$$

Thus, motion averaged over one period of the forcing term, is $X(t)$. Next, substituting (A.5) into (A.3) and using the high-frequency assumption and the smallness of α, the equation of forced oscillations reads

doi:10.1088/978-0-7503-3199-9ch6

© IOP Publishing Ltd 2023

$$m\ddot{\alpha}(t) = f(t). \tag{A.8}$$

The integration of this equation leads to

$$\alpha(t) = -\frac{f(t)}{m\omega^2}. \tag{A.9}$$

Next, taking the time-average of (A.4), produce the equation for the unperturbed trajectory,

$$m\ddot{X} = -\frac{\mathrm{d}}{\mathrm{d}X}\left(V + \frac{1}{2m\omega^2}\langle f^2(t)\rangle_T\right), \tag{A.10}$$

where

$$\langle f^2(t)\rangle_T = \frac{1}{2}(f_1^2 + f_2^2). \tag{A.11}$$

It can be therefore concluded that the average trajectory of the particle is controlled by the effective potential which comprise the time-independent potential U and an additional potential, time-independent as well, which depends on the square of the external force amplitude,

$$V_{\text{eff}}(x) = U(x) + \frac{1}{2m\omega^2}\langle f^2(t)\rangle_T. \tag{A.12}$$

The above considerations can be generalized to a charged particle in a high-frequency external field using the Kramers–Henneberger framework. In the unperturbed case the particle motion is controlled by the static potential $V(x)$ whereas after the external field is switched-on, the motion takes place in the time-dependent potential $V(x + \alpha(t))$, which is next expanded in Fourier series

$$V(x + \alpha(t)) = \sum_{n=-\infty}^{\infty} V_n(x)\mathrm{e}^{in\omega t}. \tag{A.13}$$

The Fourier coefficients are obtained by inverting (A.13)

$$V_n(x) = \frac{1}{T}\int_0^T \mathrm{e}^{-\mathrm{i}n\omega t}V(x + \alpha(t))\mathrm{d}t. \tag{A.14}$$

Following the steps outlined earlier, it can be easily concluded that the averaged particle motion takes place in the effective potential

$$V_{\text{eff}}(x) = V(x) + \frac{1}{4m\omega^2}\sum_n(F_n^2(x) + G_n^2(x)), \tag{A.15}$$

where

$$F_n(x) = V_n(x) + V_n^*(x), \qquad G_n(x) = -\mathrm{i}(V_n(x) - V_n^*(x)). \tag{A.16}$$

In many situations it is found that the sum (A.13) can be truncated to a few terms, i.e. take into account in the physical problem only a small number of exchanged photons between the particle and the electromagnetic field. Based on the assumption that the oscillations superposed on the unperturbed trajectory are of small amplitude, the time-dependent potential is expanded in powers of α,

$$V(x + \alpha(t)) = V(x) + \sum_{n\geqslant 1} \frac{\alpha(t)^n}{n!} V^{(n)}(x). \tag{A.17}$$

On the other hand, as seen above, $V(x + \alpha(t))$ can be expanded in Fourier series

$$V(x + \alpha(t)) = V(x) + \sum_{n\geqslant 1}^{\infty} (F_n(x)\cos(n\omega t) + G_n(x)\sin(n\omega t)). \tag{A.18}$$

Using the identities [2]

$$\sin^{2n}\omega t = \frac{1}{2^{2n}} \sum_{k=0}^{n-1} [(-)^{n-k} 2C_{2n}^{k} \cos 2(n-k)\omega t + C_{2n}^{n}], \tag{A.19}$$

$$\sin^{2n+1}\omega t = \frac{1}{2^{2n-2}} \sum_{k=0}^{n-1} (-)^{n+k-1} C_{2n-1}^{k} \sin(2n - 2k - 1)\omega t, \tag{A.20}$$

and choosing $\alpha(t) = \alpha_0 \sin \omega t$, the Fourier amplitudes can be expressed in terms of power series of α,

$$F_{2n}(x) = (-)^n \sum_{m\geqslant n} \frac{\alpha^{2m}}{2^{2m-1}} \frac{V^{(2m)}(x)}{(m+n)!(m-n)!}, \qquad F_{2n+1}(x) = 0 \tag{A.21}$$

$$G_{2n+1}(x) = (-)^n \sum_{m\geqslant n} \frac{\alpha^{2m+1}}{2^{2m}} \frac{V^{(2m+1)}(x)}{(m+n+1)!(m-n)!}, \qquad G_{2n}(x) = 0. \tag{A.22}$$

A.2 Bessel functions

The Bessel function of integer order [3] admits the following properties.

Infinite sum representation:

$$J_n(z) = \sum_{m}^{\infty} \frac{(-)^m \left(\frac{z}{2}\right)^{n+2m}}{m!(n+m)!}. \tag{A.23}$$

Integral representation:

$$J_n(z) = \frac{1}{2\pi} \int_{-\pi}^{\pi} e^{iz\cos\theta} e^{in\left(\theta - \frac{1}{2}\pi\right)} d\theta. \tag{A.24}$$

Recurrence relations:

$$J_{n-1}(z) - J_{n+1}(z) = 2J_n'(z), \tag{A.25}$$

$$J_{n-1}(z) + J_{n+1}(z) = 2nz^{-1}J_n(z), \tag{A.26}$$

$$J_{n+1}(z) = nz^{-1}J_n(z) - J_n'(z), \tag{A.27}$$

$$J_n'(z) = J_{n-1}(z) - nz^{-1}J_n(z). \tag{A.28}$$

Relations between solutions:

$$J_{\pm n}(\mp z) = (-)^n J_n(z). \tag{A.29}$$

Ascending series:

$$J_n(z) = \left(\frac{z}{2}\right)^n \sum_{m=0}^{\infty} \frac{(-)^m z^{2m}}{2^{2m} m!(n+m)!}. \tag{A.30}$$

The generating function:

$$\mathrm{e}^{-\mathrm{i}u \sin \omega t} = \sum_{n=-\infty}^{\infty} \mathrm{e}^{-\mathrm{i}n\omega t} J_n(u), \tag{A.31}$$

$$\mathrm{e}^{-\mathrm{i}v \cos \omega t} = \sum_{n=-\infty}^{\infty} (-\mathrm{i})^n \mathrm{e}^{-\mathrm{i}n\omega t} J_n(v) \tag{A.32}$$

and the consequences of Graf's addition theorem [3], i.e.

$$\sum_m J_{n+m}(u) J_m(v) \mathrm{e}^{\mathrm{i}m\frac{\pi}{2}} = J_n(w) \mathrm{e}^{\mathrm{i}n\Psi}, \tag{A.33}$$

where

$$w = \sqrt{u^2 + v^2}; \qquad \cos \Psi = \frac{1}{w} u; \qquad \sin \Psi = \frac{1}{w} v, \tag{A.34}$$

lead to the Jacobi–Anger expansion

$$\mathrm{e}^{-i(u \sin \omega t + \cos \omega t)} = \sum_{n=-\infty}^{\infty} J_n(w) \mathrm{e}^{-\mathrm{i}n(\omega t + \Psi)}. \tag{A.35}$$

The generalized Bessel functions (GBFs) were introduced by Reiss in [4] with the aim to handle relativistic problems in scattering theory of charged particles and radiation. The GBF $J_n(u, v)$, belongs to a set of Bessel-type functions characterized by an integer index n and two real variables (u, v). It is defined by the integral representation

$$J_n(u, v) = \frac{1}{2\pi} \int_{-\pi}^{+\pi} \mathrm{d}\theta \mathrm{e}^{\mathrm{i}(u \sin \theta + v \sin 2\theta - n\theta)}. \tag{A.36}$$

This function admits a double infinite sum representation

$$J_n(u, v) = \sum_{k=-\infty}^{+\infty} J_{n-2k}(u)J_k(v) \tag{A.37}$$

$$= J_0(u)J_{n/2}(v) + \sum_{k=1}^{+\infty} J_{2k}(u)[J_{k+n/2}(v) + J_{-k+n/2}(v)], \quad n \text{ even} \tag{A.38}$$

$$= \sum_{k=1}^{+\infty} J_{2k-1}(u)\left[J_{-k+(n+1)/2}(v) - J_{k+(n-1)/2}(v)\right], \quad n \text{ odd.} \tag{A.39}$$

From either of the above equations it follows that

$$\lim_{v\to 0} J_n(u, v) = J_n(u), \tag{A.40}$$

and

$$\lim_{u\to 0} J_n(0, v) = J_{n/2}(v), \quad n = \text{even}, \tag{A.41}$$

$$\lim_{u\to 0} J_n(0, v) = 0, \quad n = \text{odd}. \tag{A.42}$$

Symmetry relations for the GBF can be summarized by the relations

$$J_n(-u, v) = (-)^n J_n(u, v), \tag{A.43}$$

$$J_n(u, -v) = (-)^n J_{-n}(u, v). \tag{A.44}$$

Two important theorems for the GBF can be established starting from the above definition,

$$e^{i(u\sin\phi + v\sin 2\phi)} = \sum_{n=-\infty}^{\infty} e^{in\phi} J_n(u, v), \tag{A.45}$$

and

$$\sum_{k=-\infty}^{\infty} J_{n\mp k}(u, v)J_k(u', v') = J_n(u \pm u', v + v'). \tag{A.46}$$

For small arguments

$$\begin{aligned}
J_n(u, v) &\approx J_n(u) + \frac{1}{2}v[J_{n-2}(u) - J_{n+2}(u)], \quad |v| \ll 1 \\
J_n(u, v) &\approx \left[1 - \frac{u^2}{4}\left(1 - \frac{n}{2v}\right)J_{n/2}(v)\right], \quad |u| \ll 1, \quad n \text{ even} \\
J_n(u, v) &\approx \frac{1}{2}u\left[J_{(n-1)/2}(v) - J_{(n+1)/2}(v)\right], \quad |u| \ll 1, \quad n \text{ odd.}
\end{aligned} \tag{A.47}$$

Of particular interest for strong-field applications [4] is the limit $\lim_{z\to 0} J_n(z^{1/2}\mu, z\nu)$. This limit can be evaluated using the representation (A.37) of the GBF. The product $J_{2k}(u)J_{-k+n/2}(v)$ contributes terms of minimum order $n/2$ in z for $k \leqslant n/2$, thus making the function $J_n(z^{1/2}\mu, z\nu)$ behave as $z^{n/2}$ for small z,

$$\lim_{z\to 0} J_n(z^{1/2}\mu, z\nu) = \lim_{z\to 0}\left(\frac{z\nu}{2}\right)^{n/2} \sum_{k=0}^{n/2} \frac{(\mu^2/2\nu)^k}{(2k)!\left(\frac{1}{2}n - k\right)!}, \qquad n \text{ even} \tag{A.48}$$

$$= \lim_{z\to 0}\left(\frac{z\nu}{2}\right)^{n/2} \sum_{k=0}^{n/2} \frac{(\mu^2/2\nu)^{k+\frac{1}{2}}}{(2k+1)!\left(\frac{1}{2}(n-1) - k\right)!}, \qquad n \text{ odd}. \tag{A.49}$$

References

[1] Landau L and Lifshitz L 1976 *Mechanics* 3rd edn (Amsterdam: Elsevier)

[2] Gradshteyn I S and Ryzhik I M 1965 *Table of Integrals, Series and Products* 4th edn (New York: Academic)

[3] Bateman H 1953 *Higher Transcendental Functions* (New York: McGraw-Hill)

[4] Reiss H R 1980 Effect of an intense electromagnetic field on a weakly bound system *Phys. Rev.* A **22** 1786–813

www.ingramcontent.com/pod-product-compliance
Ingram Content Group UK Ltd.
Pitfield, Milton Keynes, MK11 3LW, UK
UKHW051849100926
8556IPUK00005B/539

9 780750 331975